AF531654

Plant Biotechnology
Trends and Techniques

Plant Biotechnology Trends and Techniques

Dr. Anirudh Garg

Plant Biotechnology Trends and Techniques

ISBN 978-93-5111-470-3

Published in 2015 in India by

RANDOM PUBLICATIONS

4376-A/4B, Gali Murari Lal, Ansari Road
New Delhi-110 002
Phone : +9111-43580356, 011-23289044, 011-43142548
e-mail: sales@randompublications.com,
info@randompublications.com, randomexports@gmail.com
Reprinted 2018

Type Setting by : Friends Media, Delhi-110089
Digitally Printed at : Replika Press Pvt. Ltd.

Preface

For centuries, humankind has made improvements to crop plants through selective breeding and hybridization—the controlled pollination of plants. Plant biotechnology is an extension of this traditional plant breeding with one very important difference—plant biotechnology allows for the transfer of a greater variety of genetic information in a more precise, controlled manner. Unlike traditional plant breeding, which involves the crossing of hundreds or thousands of genes, plant biotechnology allows for the transfer of only one or a few desirable genes. This more precise science allows plant breeders to develop crops with specific beneficial traits and without undesirable traits.

Many of these beneficial traits in new plant varieties fight plant pests —insects, disease and weeds—that can be devastating to crops. Others provide quality improvements, such as tastier fruits and vegetables; processing advantages, such as tomatoes with higher solids content; and nutrition enhancements, such as oil seeds that produce oils with lower saturated fat content. Crop improvements like these can help provide an abundant, healthful food supply and protect our environment for future generations.

The present book explores contemporary techniques and applications of plant biotechnology, illustrating the tremendous potential this technology has to change our world by improving the food supply. It guides students from plant biology and genetics to breeding to principles and applications of plant biotechnology. This book is recommended for junior- and senior-level courses in plant biotechnology or plant genetics and for courses devoted to special topics at both the undergraduate and graduate levels. It is also an ideal reference for practitioners.

Author

Contents

1

Plant Biotechnology: An Overview

At its simplest, biotechnology is technology based on biology - biotechnology harnesses cellular and biomolecular processes to develop technologies and products that help improve our lives and the health of our planet. We have used the biological processes of microorganisms for more than 6,000 years to make useful food products, such as bread and cheese, and to preserve dairy products.

Modern biotechnology provides breakthrough products and technologies to combat debilitating and rare diseases, reduce our environmental footprint, feed the hungry, use less and cleaner energy, and have safer, cleaner and more efficient industrial manufacturing processes.

Currently, there are more than 250 biotechnology health care products and vaccines available to patients, many for previously untreatable diseases. More than 13.3 million farmers around the world use agricultural biotechnology to increase yields, prevent damage from insects and pests and reduce farming's impact on the environment. And more than 50 biorefineries are being built across North America to test and refine technologies to produce biofuels and chemicals from renewable biomass, which can help reduce greenhouse gas emissions.

Applications of Biotechnology

Recent advances in biotechnology are helping us prepare for and meet society’s most pressing challenges. Here's how:

Heal the World

Biotech is helping to heal the world by harnessing nature's own toolbox and using our own genetic makeup to heal and guide lines of research by:

— Reducing rates of infectious disease;
— Saving millions of children's lives;
— Changing the odds of serious, life-threatening conditions affecting millions around the world;
— Tailoring treatments to individuals to minimize health risks and side effects;
— Creating more precise tools for disease detection; and

Combating serious illnesses and everyday threats confronting the developing world.

Fuel the World

Biotech uses biological processes such as fermentation and harnesses biocatalysts such as enzymes, yeast, and other microbes to become microscopic manufacturing plants. Biotech is helping to fuel the world by:

— Streamlining the steps in chemical manufacturing processes by 80% or more;
— Lowering the temperature for cleaning clothes and potentially saving $4.1 billion annually;
— Improving manufacturing process efficiency to save 50% or more on operating costs;
— Reducing use of and reliance on petrochemicals;
— Using biofuels to cut greenhouse gas emissions by 52% or more;
— Decreasing water usage and waste generation; and
— Tapping into the full potential of traditional biomass waste products.

Feed the World

Biotech improves crop insect resistance, enhances crop herbicide tolerance and facilitates the use of more environmentally sustainable farming practices. Biotech is helping to feed the world by:

— Generating higher crop yields with fewer inputs;
— Lowering volumes of agricultural chemicals required by crops-limiting the run-off of these products into the environment;

— Using biotech crops that need fewer applications of pesticides and that allow farmers to reduce tilling farmland;
— Developing crops with enhanced nutrition profiles that solve vitamin and nutrient deficiencies;
— Producing foods free of allergens and toxins such as mycotoxin; and
— Improving food and crop oil content to help improve cardiovascular health.

Biotechnology in Crop Production

Biotechnology is a word used to describe the process of using living organisms (such as plants, animals, or microbes) or any part of these organisms to create new or improved products. The term is most often associated with methods and techniques developed during the past 20 years that make use of the cells and biological molecules of living organisms. But biotechnology, in some ways, is not very new at all. For six thousand years, we have been using microbes to make useful products such as bread, yogurt, and cheese. Also, for thousands of years, people have altered the genetic makeup of plants and animals to make them more useful. For example, growers have improved the characteristics of crops by planting seeds selected from the biggest and best individual plants. Through this selection process, desirable characteristics have been promoted. Plants have been cultivated and animals have been domesticated, changing them from their wild beginnings to forms more useful to humans. While this process is slow, it has been quite successful.

Early Breeding Programs

Because people involved in agriculture could clearly see the benefits of improving plants and animals, scientists developed a keen interest in understanding how organisms inherit certain characteristics. Plant breeding as a science began in the 19th century with the discovery of how plant traits are inherited. Gregor Mendel discovered that the characteristics of plants and animals are determined by genes, which are a fundamental part of the nucleus of each cell in an organism, and that the genes carry certain traits from each parent to the offspring when organisms reproduce. Of course, not all plants and animals of a given species are identical; some have different traits that make them more desirable for agricultural production—for example, higher yield or greater resistance to drought. Plant breeders select plants with desirable traits after the exchange of genes by cross-fertilization between two parents. They then ensure that these traits will be passed on to future generations, thus creating a new variety of the plant. (A variety is a subgroup within a species and has distinctive traits that are different from other members of the species.)

Most of the early breeding programs were based on the collection of different specimens of the same species of plant from throughout the world. This approach allowed breeders to obtain as much of the genetic variation as was available within the species worldwide. Then through conventional breeding techniques they tried to utilize desirable characteristics to improve the domesticated varieties of those species.

This process is the basis for the development of essentially all varieties of plants and animals used in agriculture today. However, it is slow, commonly requiring 10 years before a new variety is ready to be released. But the biggest problem is that a desirable characteristic being sought to improve a given species may not be found among any of the plants or animals of that species in the world. Further improvement by conventional breeding can thus reach a dead end for that desired trait. For example, if resistance to a particular insect is needed in a given crop species and no such resistance is found in any plant of that species in the world, then protection of that crop may be dependent upon insecticides.

With new techniques, however, the path to genetic improvement need not end that soon. If resistance to the insect is found within a different species (animal, plant, or microbe), modern methods make it possible to transfer that resistance to the plant species of interest. The same methods could be used to increase yield, improve cold hardiness, or add any desirable characteristic. The source for genetic resources is expanded to include all living species, making the possibility of improvements virtually unlimited. Modern techniques fall into three categories: cross-breeding between species, cell fusion, and genetic engineering.

Breeding Between Species

The reproductive process through which genetic material is transferred between individuals of a species does not normally work between species. The desire to improve certain crops, however, has led plant breeders to develop processes for interspecific (between-species) hybridization to transfer genes from one plant to another plant of a related species. This transfer has been accomplished during the last 80 years only through the discovery of efficient ways to circumvent the natural barriers to the exchange of genetic information between species.

The processes for interspecific gene transfers are usually laborious, time consuming, and quite often unsuccessful. There are many steps in the process of developing a hybrid plant, and each step is a possible point for failure, either through death or sterility of the plant. Even when a hybrid plant is created, there is a possibility that undesirable traits that affect crop quality, yield, or adaptation to stress may be transferred along with the desirable trait.

Even with all of these problems, interspecific hybridization has been responsible for many improvements in the tolerance of crops to physical stresses (such as cold), resistance to diseases and insects, and increased yield potential for certain crops. The process has even made possible the development of a new grain crop, triticale, from the hybridization of wheat and rye.

Cell Fusion

Microbiologists and breeders have been looking for alternate means to transfer genes. Such exchange of genes depends on the scientist's ability to regenerate a plant from tissues or organs (leaves, stems, or even clumps of cells in culture). Much effort has been devoted to cell fusion, a process for combining two cells of the same or different species in the presence of certain chemicals or electrical current. In addition to the problem of regenerating a complete plant from the resulting cell, it is often difficult to achieve a stable hybrid (one that will retain the same characteristics through many generations) because of incompatibility between the parent species. As with interspecific hybridization, many genes are transferred during the cell fusion process, and thus unwanted traits may be transferred with the desirable ones. Further crosses may be necessary to obtain plants with the desired combination of parental traits.

Genetic Engineering

Another method for transferring genes is to move the DNA directly. (DNA is the molecule within the cell that carries genetic information.) Direct combining of DNA is called genetic engineering. New techniques make it possible to use genes from totally unrelated organisms. First, scientists identify genes that control specific functions or characteristics—for example, improved insect resistance or production of enzymes for the degradation of a certain herbicide. Then they move these genes into the host plant. *One of the advantages of this system is the ability to move just the trait of interest without carrying along any undesirable genes.*

With genetic engineering, the time from discovery to incorporation of a desirable trait into a host organism will usually be shorter than with the use of classical breeding techniques to produce new varieties.

Genetic engineering may help crop producers reduce dependence on pesticides by affecting the plant's resistance or tolerance to pests in the field. It can also affect production efficiency and profitability by improving the climatic adaptation of the plant (for example, its tolerance to cold or drought) and by enhancing the postharvest storability of the product.

Modern Plant Biotechnology

Today, biotechnology is being used as a tool to give plants new traits that benefit agricultural production, the environment, and human nutrition and health. The goal of plant breeding is to combine desirable traits from different varieties of plants to produce plants of superior quality. This approach to improving crop production has been very successful over the years.

For example, it would be beneficial to cross a tomato plant that bears sweeter fruit with one that exhibits increased disease resistance. To do this, it takes many years of crossing and backcrossing generations of plants to obtain the desired trait. Along the way, undesirable traits may be manifested in the plants because there is no way to select for one trait without affecting others. Another limitation of traditional plant selection is that breeding is restricted to plants that can sexually mate. Advances in scientific discovery and laboratory techniques during the last half of the twentieth century led to the ability to manipulate the deoxyribonucleic acid (DNA) of organisms, which accelerated the process of plant improvement through the use of biotechnology.

Genes and the Genome

Plants are made of millions of cells all working together. Every cell of a plant has a complete "instruction manual" or genome that is inherited from the parents of the plant as a combination of their genomes. Genes are found within the genome and serve as the "words" of the instruction manual. When a cell reads a word, or in scientific terms "expresses a gene," a specific protein is produced. Proteins give an individual cell, and therefore the plant, its form and function. Genes (words) are written using the four-letter alphabet A, C, G, T. The letters are abbreviations for four chemicals called bases, which together make up DNA. DNA is universal in nature, meaning that the four chemical bases of DNA are the same in all living organisms. Consequently, a gene from one organism can function in any other organism.

The ability to move genes into plants from other organisms, thereby producing new proteins in the plant, has resulted in significant achievements in plant biotechnology that were not possible using traditional breeding practices.

Genes into Plants

To genetically modify a plant, the thousands of bases of DNA comprising an individual gene are transferred into an individual plant cell where the new gene becomes a permanent part of the cell's genome. This process makes the resulting plant "transgenic." Transfer of DNA into plant cells is done using various "transformation" techniques that are the result of discoveries in basic science.

Nature's way

One method to transfer DNA into plants takes advantage of a system found in nature. The bacterium that causes "crown gall tumors" injects its DNA into a plant genome, forcing the plant to create a suitable environment for the bacterium to live. After discovering this process, scientists were able to "disarm" the bacterium, put new genes into it, and use the bacterium to harmlessly insert the desired genes into the plant genome.

Cellular target practice

In the "biolistic" or "gene gun" method, microscopic gold beads are coated with the gene of interest and shot into the plant cell with a burst of helium. Once inside the cell, the gene comes off the bead and integrates into the cell's genome.

That's shocking!

It was also discovered that plant cells could be "electroporated" or mixed with a gene and "shocked" with a pulse of electricity, causing holes to form in the cell through which the DNA could flow. The cell is subsequently able to repair the holes and the gene becomes a part of the plant genome.

Selecting the right cells

When using these methods, new genes are successfully introduced into only a small percentage of the cells, so scientists must be able to "pick out" or "select" the transformed cells before proceeding. This is often done by concurrently introducing an additional gene into the cell that will make it resistant to an antibiotic. A cell that survives antibiotic treatment will most likely have received the gene of interest as well; that cell is subsequently used to propagate the new plant. There is a concern that the gene giving antibiotic resistance could naturally be transferred to bacteria once the transgenic plant is in the wild, making bacteria resistant to antibiotics that are used to fight human infection. Scientists are currently devising ways to select for transformed cells that will alleviate this issue.

Traits in Plants

Changes made to plants through the use of biotech- nology can be categorized into the three broad areas of input, output, and value-added traits. Examples of each are described below.

Input traits

An "input" trait helps producers by lowering the cost of production, improving crop yields, and reducing the level of chemicals required for the control of insects, diseases, and weeds.

Input traits that are commercially available or being tested in plants:

- Resistance to destruction by insects
- Tolerance to broad-spectrum herbicides
- Resistance to diseases caused by viruses, bacteria, fungi, and worms
- Protection from environmental stresses such as heat, cold, drought, and high salt concentration

Output traits

An "output" trait helps consumers by enhancing the quality of the food and fiber products they use. Output traits that consumers may one day be able to take advantage of:

— Nutritionally enhanced foods that contain more starch or protein, more vitamins, more anti-oxidants (to reduce the risk of certain cancers), and fewer trans-fatty acids (to lower the risk of heart disease)

— Foods with improved taste, increased shelf-life, and better ripening characteristics

— Trees that make it possible to produce paper with less environmental damage

— Nicotine-free tobacco

— Ornamental flowers with new colors, fragrances, and increased longevity

Value-added" traits

Genes are being placed into plants that completely change the way they are used. Plants may be used as "manufacturing facilities" to inexpensively produce large quantities of materials including:

— Therapeutic proteins for disease treatment and vaccination

— Textile fibers

— Biodegradable plastics

— Oils for use in paints, detergents, and lubricants

Plants are being produced with entirely new functions that enable them to do things such as:

— Detect and/or dispose of environmental contaminants like mercury, lead, and petroleum products

BIOTECHNOLOGY FOR PEST MANAGEMENT

Crop breeders have traditionally considered the development of pest resistance in crops and animals to be of primary importance. Genetic engineering can speed up the process of developing resistant varieties and potentially can introduce genes that will provide new ways for plants to withstand pests.

Insect Control

There has been much interest in using genetic engineering for insect control. The interest in developing insect-resistant varieties results from the desire to reduce the extent of pesticide use and to avoid the development of pesticide resistance in some of the important insect pests. The development of crops with insect resistance is not a new idea. Classical breeding programs have been developing varieties with insect and disease resistance or tolerance for many years. The problem has been that insects are generally able to overcome the resistance in 2 to 10 years.

Another aim of biotechnology in insect control is to provide materials that are selective, will not affect nontarget species (including humans), and to which insects will not easily develop resistance. Much of the work to date has focused on applications of *Bacillus thuringiensis* (Bt), an insect control bacterium, but other approaches are being explored as well.

Recent research on Bt has followed two approaches. The first is the selection and development of Bt strains that are specific for other pests or that produce higher concentrations of the compounds. Different strains of Bt to control plant- and animal parasitic nematodes, animal-parasitic liver flukes, protozoan pathogens, and mites have been identified.

The second approach is to move the gene that controls the production of the Bt proteins into crop plants. Bt genes for control of an insect species have been incorporated into tomatoes, tobacco, cotton, and corn with good results. Bt proteins are highly specific to particular insect species. Therefore, to achieve control of several different insect species at one time, many different gene codes would have to be incorporated into the plant's genetic makeup.

The incorporation of Bt genes into plants causes concern that insects will likely develop resistance to the Bt strain. The probability is especially high in current cases where the gene is expressed throughout the plant. For example, if three to four generations of an insect feed on the plant leaves or fruits throughout the growing season, the insect would be constantly exposed to the chemical. It is likely that the insect will develop resistance to the chemical. It may become possible to direct the expression of the Bt products to only the fruit tissues, meaning that only one generation of insects would be exposed. It may also become possible to incorporate the genes in such a way that they are expressed only after an insect begins to feed on the plant. Finally, several different Bt genes may be introduced at one time, making the development of resistance more difficult.

Other Approaches to Insect Control. Research has been conducted on developing insect resistance mechanisms that can be used in different plant species. This work has centered on a chemical complex known as the cowpea trypsin inhibitor (CpTI). Results obtained so far suggest that CpTI will control a wider range of insects than the specific Bt products. In experiments on tobacco, good control of foliage feeding caterpillars was achieved. There appears to be no adverse affect on humans because CpTI comes from the cowpea and has not appeared to cause health problems when cowpeas are eaten raw or cooked.

Another area of interest is the incorporation of insecticide resistance into the natural predators or parasites of major and secondary insect pests. Secondary insect pests are those that are controlled by natural predators until insecticides are applied to control a major insect pest. The insecticides also kill the natural predators, allowing the secondary insect pests to flourish. Resistant types of the predators and parasites are being sought so they can be used for control. Modern genetic techniques will help in understanding the mechanisms of insect resistance and in the development of greater resistance in the future.

Weed Control

Herbicide Tolerance. One of the first commercial uses of biotechnology involves the genetic improvement of crop tolerance to herbicides. *Herbicide tolerance* is a plant's ability to endure the effects of a herbicide at the rate normally used in agricultural production. *Herbicide resistance* is the ability of a plant to be unaffected at any feasible rate of herbicide application. Most crops are resistant to one or more herbicides. For example, corn is naturally resistant to atrazine, corn and soybeans are tolerant to alachlor (Lasso) and metachlor (Dual), but soybeans are not tolerant to atrazine.

Biotechnology has provided plant scientists with additional tools to determine the chemical and genetic modes of action of many of these herbicides and also the mechanisms that account for a plant's natural tolerance or resistance to herbicides. As a result, scientists will use this knowledge to incorporate herbicide tolerance into crop plant species.

Several different methods have been used successfully to develop herbicide-resistant crop varieties. One method is to find a closely related species that has herbicide tolerance or resistance and then, through classical breeding techniques, to incorporate that tolerance into the desired plant. This process has been successfully applied to canola (oilseed rape) using a related species (bird's rape) that was found to be resistant to atrazine. This new atrazine-resistant variety of canola is currently being cultivated in Canada.

A second method has been the use of cell or tissue culture to test many different lines of plants for tolerance to a specific herbicide. As a result, Pioneer Hi-Bred has been able to develop three corn hybrids for use with Pursuit, a herbicide that typically kills conventional corn hybrids.

A third method has been to determine the specific gene or genes within a plant or microbe that allow tolerance or resistance to a specific herbicide. This gene is inserted into the plant of interest, which is then tested for tolerance to the herbicide. This method has made it possible to develop cotton tolerant to the herbicide bromoxynil (Buctril) and to soybeans tolerant of glyphosate (Roundup).

Bioherbicides. In the near future, biotechnology will probably influence weed control in at least four ways. The first is through the production of *bioherbicides.* Bioherbicides are fungal or bacterial products selected for their ability to cause disease in specific plants, such as weeds, without harming desirable plants. They may be applied in the same manner as conventional herbicides.

At present, two bioherbicides are being marketed for the control of specific weeds that are normally hard to control: DeVine for control of strangler vine in Florida citrus and Collego for control of northern jointvetch in rice and soybeans in Arkansas, Louisiana, and Mississippi. DeVine has been so successful in destroying strangler vine that the market for the product has almost been lost. The reason for its great effectiveness is that the product remains in the soil and gives 95 to 100 percent control for 6 to 10 years after a single application.

There are other bioherbicides in various stages of research and development for such things as control of prickly sida in cotton and soybeans, control of sicklepod in cotton and soybeans, control of spurred anode in cotton, control of velvetleaf, and growth suppression of water hyacinth. Biotechnology will play a major role in helping overcome problems in manufacturing these bioherbicides by the development of better fermentation processes, as well as assisting in the isolation of the genetic determinants of virulence, specificity, sporulation capacity, toxin production, and tolerance to climatic stresses.

Natural Control Compounds. A second application of biotechnology that shows promise is the development of microbial and secondary plant products for use as herbicides. Much effort has been devoted to determining the actual compounds associated with allelopathy (the ability of one plant to influence the growth of others nearby by releasing chemical compounds). Many of these compounds have limited selectivity and a lack of stability. However, one herbicide (Herbiaceae) derived from chemicals found in a naturally occurring microbe is being marketed in Japan. Herbiaceae exhibits strong herbicidal activity against a wide spectrum of grass and broadleaf weeds when it is applied to foliage.

A more important role for these compounds is to provide models for the development of new chemicals that could be produced as commercial herbicides. It may be possible to produce synthetic derivatives of these chemicals that are more stable under field conditions, have greater selectivity than the natural chemicals, or have other advantages over the original chemicals. Several companies have developed chemicals based on this natural herbicide chemistry. It is believed that the naturally occurring herbicides will be safer for the environment because many of them are degraded rapidly in the soil.

Potential Concerns. Concerns have been raised about these developments. One is that development of herbicide-resistant or herbicide-tolerant crop varieties may lead to overdependence on herbicides. Many of the herbicides being used for the development of crop herbicide resistance are broad spectrum, low-mammalian-toxicity chemicals that are thought to be safer for the environment than conventional herbicides. However, attempts are also being made to develop varieties with resistance to certain other compounds, such as atrazine, that are more persistent and have been found in groundwater in certain areas of the United States. Continued use of these older chemicals, ones that tend to persist in the soil and may move into groundwater, could cause environmental problems.

Associated with the exclusive use of one herbicide for weed control in each crop in a rotation is the possible development of herbicide-resistant weeds. There are already over 100 species of weed plants known to have developed resistance to one herbicide or another. Much of this resistance has resulted from the continual use of one type of herbicide (for example, triazines) increasing the probability of developing resistant plants within the weed population. The use of a comprehensive weed control program, including rotating chemicals based on their mode of action, should allow the successful use of the new, "safer" herbicides.

Another concern is the possible movement (outcrossing) of the "engineered" genes from the host plant to related weed species. This possibility has not received much attention in the United States because in most cases our weed species are not closely related to crop species. An exception is the vegetable industry in California because many of the vegetable crops have closely related weed relatives in the wild. On the other hand, the possibility of outcrossing may not be very likely, considering the lengths to which plant breeders had to go to accomplish interspecific hybridisation before genetic engineering was introduced.

Disease Control

Control of disease is a subject of great interest for biotechnologists. The majority of advances have been in control of viral diseases. Because most viruses are spread mechanically or through insect vectors, control efforts have traditionally revolved around control of the vectors and destruction of diseased plant material.

Viruses are composed of two parts—the viral DNA and a coat protein that surrounds the viral DNA. Researchers have known about the phenomenon of cross-protection: that infection of a plant by a mild strain of virus can often protect the plant from a serious infection by a more virulent related strain. The researchers have recently discovered that it is the presence of the coat protein that restricts the infection by the virus in cross-protection. By incorporating the genes for the coat protein into the plant it is possible to have the plant itself produce low levels of the coat protein. These low levels of the coat protein delay or restrict infection of the plant by the virus.

An example is the incorporation of genes for the production of the coat protein of the tobacco mosaic virus (TMV) into tomato plants. Untreated tomato plants infested with TMV showed up to 60 percent loss in yield, whereas resistant plants showed no yield decrease after inoculation with the virus. The level of the viral coat protein in transgenic plants (those that contain genetic material from a different organism) is lower than that found in plants infected with endemic strains of the virus.

Cross-protection via gene transfer offers a number of advantages. The first is that the protection is essentially permanent, similar to that afforded animals by vaccines. The need for the use of chemicals to control insect vectors is also reduced.

Other possible aspects of disease control in which biotechnology may play a role include the use of plant disease biocontrol fungi and the development of fungal resistance in plants. Plant disease biocontrol fungi are naturally occurring organisms that are antagonists for certain soil-borne plant pathogens. Biotechnology will play a role in better understanding the mechanisms for host specificity and virulence of these fungi and may lead to the production and formulation of control products.

Fungal resistance is an area of interest, but no great progress has been made yet. Biotechnology will enhance our understanding of the mechanisms that control a plant's ability to recognise and defend itself against disease-causing fungi.

Biotechnology and Improvement of Product Quality

Although biotechnology efforts have focused primarily on pest management, there are other potential applications to crop production. The possibilities are highly varied, and new applications are developing so quickly that it is difficult to keep up with all of them. Here are a few examples of current biotechnology research on plants.

Value of Plant Products

Plant breeders have long been interested in modifying components of crop products, such as the amount of various amino acids contained in the proteins of corn used for human and animal feeds. Traditional breeding has developed cultivars with high concentrations of lysine, an essential amino acid. The increase in lysine content, however, was accompanied by a 10 percent decrease in yield. Scientists are working to incorporate a gene in corn that will increase the amount of lysine without the associated yield decrease. They have also developed corn with a higher-than-usual oil content for use as a source of vegetable oil and feed.

Oilseeds are another crop of interest. Researchers are seeking to improve the nutritional qualities of vegetable oils along with the type and concentration of these oils within the plant seeds. Work is under way to develop varieties of sunflowers and canola with high oleic acid content. Researchers have also produced a low-palmitic-acid soybean, making the oil lower in saturated fats and therefore more comparable nutritionally to canola oil. Bioengineers have recently been able to alter canola so that it can also produce lauric acid, a key raw material for the soap, detergent, oleochemical, personal care, and food industries. Currently, the only commercial sources of lauric acid are coconut and palm kernel oils.

Improving Quality

Food producers and processors are interested in the ripening of fruit and in the processing quality of fruits and vegetables. An example is improving the processing quality of tomatoes, a major food crop and a significant source of vitamin C. The characteristics of most interest to processors are percentage of solids and consistency. It has been estimated that an increase of 1 percent in tomato solids would save the processing industry between $70 and $100 million annually. Another objective of tomato research is the desire to produce a better-tasting "vine-ripe" tomato that has a greater shelf life and will resist bruising during shipment. Currently, supermarket tomatoes are picked green, then artificially ripened just before shipment. This process has ensured that the tomatoes resist bruising, but flavour is affected.

Studies have shown that two different mechanisms within the tomato appear to control the ripening process. The first is the production of ethylene within the fruit. Researchers have been able to isolate the compound responsible for ethylene production and to regulate this compound, lengthening the ripening period.

The researchers found an enzyme that is responsible for degrading tomato flavour and fruit firmness. Genetic engineering techniques have allowed researchers to block this enzyme, which has led to a tomato that can remain on the vine to ripen (so that it will have better flavour) and still can be shipped with a minimum of bruising and spoilage. Processed juice made from these tomatoes has shown greater consistency and a higher percentage of solids than that made from the conventional varieties.

Advantages of Plant Conservation Genetics

Plant conservation genetics is an emerging area of science. Plant conservation genetics encompasses the analysis of wild populations of rare or endangered plants to understand the factors influencing the sustainability of the diversity within the species and of the species itself. Wider definitions include any study of plant genetics that could improve the understanding of the relationships between plants either wild or cultivated because this information is almost always likely to be useful in conservation of genetic diversity.

Conservation of plants in situ is the main option for plants with little value as ornamentals or in agriculture or forestry. Ex situ conservation is largely employed for economic species, especially those used in agriculture.

However, the conservation of many economic species can be enhanced by conservation of wild relatives in situ. Plant conservation genetics, along with most other areas of biology, has been greatly advanced by developments in molecular biology. The power of molecular genetic analysis and the increasing availability of cost-effective technologies allow plant conservation genetics to be applied much more widely in support of plant biodiversity conservation.

Plants are biologically essential to life on earth. Plant conservation underpins the maintenance of animal life. Plants provide directly or indirectly basic necessities such as food and shelter for a wide range of organisms, including humans. Plant conservation genetics is key to the management of plant genetic resources for sustainable agriculture, forestry, and food production. Humans also value plant conservation for less obvious or immediate reasons. Plant diversity contributes both directly and indirectly to biodiversity in general.

For many, the ability to experience the diversity of life forms is an important component of their quality of life. The contribution of plant conservation to the enrichment of life experiences is the primary motivation for the application of plant conservation genetics in nature conservation. A few plant species, such as the cereals, are the main components of human diets. A much larger number are minor sources of food. Many more species used in construction, clothing, and medicine are essential to many human communities.

Modern genetics builds on the understanding of inheritance as deduced by Mendel more than a century ago. It is empowered by knowledge of these mechanisms at the chemical level based on the double-helical structure of DNA elucidated over 50 years ago. Population genetics has become a well-developed discipline. Analysis of plant genomes at the DNA level allows investigation of genetic relationships between plants. Advances in techniques have made this approach much more practical in the past decade. The more recent development of genomics (the study of all the genes in organisms) promises to greatly enhance the future of plant conservation genetics.

The applications of genetic analysis in plants are manifold, including evolutionary biology, population genetics, and reproductive biology. Practical applications involve domestication and determination of gene function; determination of weediness in plant populations; and development of management strategies for conserving rare species, breeding plants, and identifying new uses for plants and plant products.

Strategies for Plant Conservation

Plant genetic resources can be conserved in situ in wild populations or ex situ in collections such as seed banks. In situ conservation is essential for the survival of a very large proportion of plant species. The use of protected areas such as reserves and national parks needs to complement strategies for the native vegetation on other public and private lands. Technical advances have enabled important gains in the ex situ conservation of plants. The ability to store seeds in seed banks is one. Maintaining living collections such as those in botanic gardens is another.

A more recent advance is the DNA bank. These banks do not at present allow recovery of entire plants but do effectively conserve the genes themselves. The choice of conservation method will depend on the species and populations under study. For example, populations conserved in situ may be vulnerable to changes in land use but should remain in situ for continued evolution of the species long term. The size of the population and the security of the site are key determinants.

Application of Plant Conservation Genetics

Genetic analyses of wild populations support their conservation by providing valuable input into management of in situ populations. This information may also assist in identifying appropriate material for ex situ conservation collections. Analysis of such material also supports resource management by reducing duplication, ensuring protection of maximum diversity, and improving the plants. Genetic analyses provide the same benefits for domesticated plants as well. To support plant diversity in wild plant populations, phylogenetic analysis based upon DNAsequencing defines evolutionary relationships between plants, and molecular analysis determines the genetic structures of plant populations. The extent of seed dispersal and pollen flow can be measured and the reproductive strategy of the plant established. These methods provide a basis for decisions to ensure conservation of a population that is viable in an evolutionary sense.

Populations of sufficient size containing any genetically unique subpopulations can be protected when appropriate data are available. Other methods support in situ and ex situ plant genetic resources as a broad base for crop production and forestry. These are focused at the genotype or gene level. Conservation of environmentally adapted genotypes provides healthy genetic backgrounds for plant breeding. Conservation of specific genes or alleles may allow enhancement of these genotypes in plant breeding for specific attributes.

Plant genome analysis addresses some essential biological questions. Comparative genome analysis yields new insights into evolutionary biology. Phylogenetic analysis based on genomics allows investigations from the level of individual gene evolution to chromosomal rearrangements. Rapid evolutionary processes may be important in understanding ecology and may operate over very small geographical ranges in response to environmental gradients. Plant conservation genetics is central to all of these issues and is crucial to extending plant biodiversity analysis beyond species richness assessments to include variation within species.

This section outlines plant conservation genetics for both ex situ and in situ approaches. Topics include plant genetic resource collections, botanic gardens, herbaria, DNA banks, the impact of habitat fragmentation, conservation of rare species, and molecular and genomic techniques. The emphasis is on the practical aspects of the strategies and the technologies for conserving plant biodiversity.

From the perspective of the practical plant conservationist, this volume describes the role of in situ and ex situ conservation in seed banks, botanic gardens, herbaria, and DNA banks and highlights developments in several underlying technologies. For example, plant phylogenetic and population genetic analyses have become more effective because of the increasingly powerful tools being developed for molecular genetic analyses. Complex legal and ethical issues shape the policies and procedures of organizations and individuals involved in plant conservation.

Humans are therefore consumers who directly or indirectly depend on plants for their food. Before humans first practiced agriculture, ancient hunter-gatherers had evolved a complex relationship with their environment. They had an intimate knowledge of the plants and animals in their surroundings and used a wide variety of plant and animal foods. Aboriginal peoples relied largely on plants, but most had diets that included animal products. With the development of agriculture some 10,000 years ago, people narrowed their food selections. Instead of the many plant species once gathered in the wild, 24 cultivated plants now account for much of the food people eat. Three of them-the cereals: wheat, rice, and maize-make up about two thirds of the human diet.

Various cultures rely on different plants as their main food crop, or staple. The major food plants evolved at the same time as the societies that use them. Thus the Japanese have many different soybean-based foods and sauces, and Westerners eat wheat under many guises. Many Latin-Americans and Africans eat cassava, a tuber crop that is virtually unknown in North America and Europe.

Some societies eat almost entirely plants, whereas others use a substantial amount of animal-derived foods. For example, in India plants supply 80% of dietary protein (cereals and legumes), but in the United States plants provide only 25% of dietary protein. The rest comes from animals and animal products. There are also differences among regions, social classes, and people of various religions. In developing countries, city dwellers typically eat more meat-historically regarded as a sign of affluence-than do farmers. For political reasons, governments often encourage meat consumption by providing agricultural price supports for feed grains, tax incentives for feedlot operators, guaranteed minimum prices, or government storage of surpluses. Researchers estimate that 85% of the calories and 80% of the protein in the human diet now come directly from plants. However, this situation 3.1 is changing as people become more affluent. Rising affluence therefore puts additional pressures on the food system.

People want to eat a more varied diet, and they want and can afford to eat more animal products. Global meat production has increased dramatically in the last 50 years, quadrupling since 1950. Until about 1950 in the industrialized countries, and even today in many parts of the developing world, farmers who practiced a sustainable mode of farming integrated livestock rearing with food crop production. They rotated food crops (wheat, potatoes, and sugar beets) with feed crops (hay, clover, and alfalfa). People ate the former; farm animals ate the latter. Farmers spread animal waste (manure) over the soil as fertilizer, and leguminous feed crops (clover and alfalfa) added nitrogen to the soil.

Modern agriculture depends on purchased inputs. Historically, the transition from stages 1 to 3 in the preceding schema has been gradual, as populations grew slowly and the necessary technologies evolved over time. The purchased inputs gradually replaced farm-produced inputs such as hay or manure. Farmers now purchase the following inputs:

Farm machinery. The process of mechanization started in the mid-19th century. New farming techniques depended on scientific advances and the industrial production of inputs. Manual labor and animal power were replaced by steam power and later by the internal combustion engine. Farmers used tractors to pull plows or the reaper-binder, which they later replaced with the combine harvester. Milking machines, cotton gins, cotton pickers, sugar beet and potato harvesters, and tomato-harvesting machines all have greatly reduced the need for manual labor. In developing countries small tractors have revolutionized agricultural production and greatly reduced the input of labor.

New and Improved Varieties. Initially, farmers selected suitable varieties of crops for planting the next year. These selected strains were the mainstay of agriculture until the science of genetics emerged in the 20th century. Then it became possible to deliberately interbreed different strains of a crop to consolidate desirable characteristics in a single strain. The first step is always to evaluate the field performance of known varieties. Scientists have systematically applied plant-breeding principles to improving rice and wheat, and have widely introduced new varieties in Latin America and Asia, displacing local varieties.

This process, often called the Green Revolution, has significantly raised crop yields in an era when the amount of land cultivated has remained more or less constant. The introduction of hybrid rice in China and Southeast Asia is raising rice yields even further. About roughly 40% of all increases in crop productivity during the past 50 years stemmed from breeding new varieties. The other 60% improvement has come from managing the crop environment by inputs such as energy, fertilizer, and pesticides.

Originally European farmers relied on manure and crop rotation to maintain fertility. Later they found that ground bones and rock phosphate enhanced crop production, as well as nitrates (long imported from Chile, in the form of fossil guano). The invention of a chemical process that combines nitrogen gas with hydrogen gas to form ammonia, allowed the widespread production of nitrogen fertilizers. This led to a tremendous increase in nitrogen fertilizer use worldwide. Fertilizer applications have been slowly decreasing in developed countries, but in developing countries they are still rising.

Herbicides and Pesticides. Herbicides replaced manual labor for weeding, and farmers used insecticides and fungicides to minimize crop losses. These changes in food production had a great impact on commercial grain farming (for example, in the United States, Canada, and Australia), on mixed crop and livestock farming (in Europe), and on intensive wetland rice farming in Asia and Africa. More complex approaches to weed control and pest control are now replacing heavy use of chemicals in agriculture.

Irrigation Technology. Irrigation has become a very important agricultural input in the past 40 years and accounts for much success in raising food production. Indeed, although only 18% of the world's arable land is irrigated, this land produces 40% of our food. Irrigated land is highly productive. However, this trend cannot be maintained: There simply is not enough water in many areas. Experts now see that water is the resource that will be most limiting for food production in the 21st century.

Information Technology. New innovations in agriculture that use information technology are often called precision agriculture. To obtain maximal yields, farmers of very large farms use remote sensing of their land and their crops by satellite or airplanes to adjust irrigation water, fertilizer application, and genetic varieties. Large tractors equipped with computers can now control row spacing and crop planting densities. Inputs do not exist in isolation, but interact. Scientists often breed improved varieties to fit other production-enhancing inputs. For example, after scientists introduced hybrid maize in the United States in the 1930s and nitrogen fertilizers became widely available, agronomists bred maize to respond to nitrogen fertilizers. After the invention of the automatic tomato harvester, tomatoes were bred to withstand the rougher handling to which such machines subject these fruit. These technological advances have led to a style of agriculture that depends on purchased inputs. A survey of U.S. production methods for maize from 1910 to the present clearly shows this dependence. The technological advances resulted from government-sponsored and industrial research, the main thrust of which has always been to develop technological inputs that let farmers minimize per unit production costs. Many of these advances diminished work opportunities on the farm in favor of jobs in towns and cities where the inputs are manufactured.

Regulation of Plant Biotechnology

Biotechnology products are regulated primarily by three federal authorities: the Food and Drug Administration (FDA), the Environmental Protection Agency (EPA), and the U.S. Department of Agriculture (USDA). The FDA has jurisdiction over new human and animal drugs as well as older drugs produced in new ways. The FDA also has jurisdiction in ensuring the safety of new foods and of new or increased amounts of substances in foods or food additives.

The EPA has jurisdiction over any product that may have pesticidal properties and for any new chemical substance for introduction into the U.S. market that is not regulated under any other statutory authority. The USDA has broad statutory powers to regulate agricultural research and agricultural products to protect crops and livestock from pests, disease, or harmful plants. Along with the federal regulation, several states, including North Carolina, have also developed legislation to control certain aspects of biotechnology.

North Carolina was one of the first states to formulate and enact legislation specifically regulating biotechnology. The North Carolina Genetically Engineered Organisms Act regulates the release into the environment and the commercial use of genetically engineered organisms. (The federal agencies listed above also regulate these activities.) The state law, administered by the North Carolina Department of Agriculture (NCDA), requires that a permit be obtained from the NCDA for field testing of genetically engineered microbes, plants, or animals.

Organisms produced by traditional breeding methods, such as hand pollination of crop plants and artificial insemination of animals, do not require a permit. The law and regulations are formulated to work with existing federal regulatory procedures, and the NCDA cooperates with federal regulatory agencies in the permit review process. The legislation regulating biotechnology must address the concerns of the general public, farmers, and the business community.

Biotechnology offers increased opportunities for improving or developing food production systems, especially from an environmental standpoint. Moreover, the potential benefits extend to improving the nutritional quality, safety, flavour, convenience, and cost of the food supply. However, acceptance by both regulatory agencies and consumers will depend on ensuring that foods resulting from the application of biotechnology are indeed safe to eat. Any changes in the composition of foods or in the methods used to handle, process, preserve, and distribute them must be evaluated to determine their impact on food safety.

The U.S. Food and Drug Administration, in late May of 1992, announced its policy for foods derived from genetically engineered plant varieties. In essence, the policy statement reaffirmed that genetically engineered foods will be judged on the characteristics of the food and not on how the plant genes may have been manipulated. The policy also stated that the anticipated regulatory approach will be "identical in principle" to that applied to foods - genetically modified by traditional plant breeding practices. Thus, it appears that the FDA views the safety considerations for genetically engineered products to be no greater than for products genetically modified by traditional practices.

The FDA's policy on the regulation of genetically engineered foods should not be viewed as fixed. On the contrary, as we expand our understanding of this new tool, policies governing its application will undoubtedly be modified. Just as policies change to reflect new understanding, so, too, can consumer attitudes. For those who are skeptical, even fearful, we hope that wise and prudent application of this technology will result in benefits that diminish these concerns.

References

Ammann, K., *Biodiversity and Agricultural Biotechnology: A Review of the Impact of Agricultural Biotechnology on Biodiversity*, Bern, Switzerland: University of Bern Botanical Garden, 2003.

Conway, G., *The Doubly Green Revolution: Food for All in the Twenty-first Century*, Ithaca: Cornell University Press, 1997.

Cowan, C. W., and P. J. Watson, eds., *The Origins of Agriculture: An International Perspective*, Washington, D.C.: Smithsonian Institution Press, 1992.

Custers, R., ed., *Safety of Genetically Engineered Crops*, Flanders: Inter-university Institute for Biotechnology, VIB publication, 2001.

Kloppenburg, J. R., Jr., *First the Seed: The Political Economy of Plant Biotechnology*, Cambridge: Cambridge University Press, 1988.

2

Plant Cell and Tissue Culture Techniques

An important aspect of all biotechnology processes is the culture of either the plant cells or animal cells or microorganisms. The cells in culture can be used for recombinant DNA technology, genetic manipulations etc.

Plant cell culture is based on the unique property of the cell-totipotency. cell-totipotency is the ability of the plant cell to regenerate into whole plant. This property of the plant cells has been exploited to regenerate plant cells under the laboratory conditions using artificial nutrient mediums. With the advances made in genetic engineering, it became possible to introduce foreign genes into cell and tissue culture systems. This led to the development of genetically modified (GM) or transgenic crops which had improved traits and characteristics.

History of Cell Culture

In the early 19th century, Schleiden and Schwann proposed the concept of the 'cell theory'. In 1902, Gottlieb Haberlandt, the german botanist and regarded as the father of plant tissue culture, first attempted to cultivate the mechanically isolated plant leaf cells on a simple nutrient medium. He did not succeed in achieving the growth and differentiation of the cultured cells, however, he predicted the concept of growth hormones, the use of embryo sac fluids, the cultivation of artificial embryos from somatic cells, etc.

During the period 1902 - 1930, attempts were made to culture the isolated plant organs such as roots and shoot apices (organ culture). Hanning (1904) isolated embryos of some crucifers and successfully grew on mineral salts and sugar solutions. Simon (1908) successfully regenerated a bulky callus, buds, roots from a poplar tree on the surface of medium containing IAA which proliferated cell division. Gautheret, White and Nobecourt (1934-1940) largely contributed to the developments made in plant tissue culture. White (1939) cultured tobacco tumour tissue from the hybrid Nicotiana glauca, and N. Langsdorffii.

The period of 1940 - 1970s saw the development of suitable nutrient media to culture plant tissues, embryos, anthers, pollen, cells and protoplasts, and the regeneration of complete plants (in vitro morphogenesis) from cultured tissues and cells. In 1941, van Overbek and co-workers used coconut milk (embryo sac fluid) for embryo development and callus formation in Datura. Steward and Reinert (1959) first discovered somatic embryo production in vitro. Maheswari and Guha (1964) developed the anther culture for the production of haplid plants. Skoog and Miller (1957) advanced the hypothesis of organogenesis in cultured callus by varying the ratio of auxin and cytokinin in the growth medium. Muir (1953) developed a successful technique for the culture of single isolated cells wich is commonly known as paper-raft nurse technique (placing a single cell on filter paper kept on an actively growing nurse tissue). In 1952, the Pfizer Inc., New York (U.S.A) got the US patent and started producing industrially the secondary metabolites of plants. The first commercial production of a natural product shikonin by cell suspension culture was obtained.

In 1980s using Genetic engineering, for the first time, it was possible to introduce foreign genes into cell and tissue culture systems to develop plants with improved characteristics (transgenic crops) which may contribute to the path towards the second green revolution.

Concepts of Plasticity and Totipotency

Two concepts, plasticity and totipotency, are central to understanding plant cell culture and regeneration. Plants, due to their sessile nature and long life span, have developed a greater ability to endure extreme conditions and predation than have animals. Many of the processes involved in plant growth and development adapt to environmental conditions. This plasticity allows plants to alter their metabolism, growth and development to best suit their environment. Particularly important aspects of this adaptation, as far as plant tissue culture and regeneration are concerned, are the abilities to initiate cell division from almost any tissue of the plant and to regenerate lost organs or undergo different developmental pathways in response to particular stimuli.

When plant cells and tissues are cultured in vitro they generally exhibit a very high degree of plasticity, which allows one type of tissue or organ to be initiated from another type. In this way, whole plants can be subsequently regenerated. This regeneration of whole organisms depends upon the concept that all plant cells can, given the correct stimuli, express the total genetic potential of the parent plant. This maintenance of genetic potential is called 'totipotency'. Plant cell culture and regeneration do, in fact, provide the most compelling evidence for totipotency. In practical terms though, identifying the culture conditions and stimuli required to manifest this totipotency can be extremely difficult and it is still a largely empirical process.

In Vitro Cultivation of Plant

When cultured in vitro, all the needs, both chemical (Table 1) and physical, of the plant cells have to met by the culture vessel, the growth medium and the external environment (light, temperature, etc.). The growth medium has to supply all the essential mineral ions required for growth and development. In many cases (as the biosynthetic capability of cells cultured in vitro may not replicate that of the parent plant), it must also supply additional organic supplements such as amino acids and vitamins. Many plant cell cultures, as they are not photosynthetic, also require the addition of a fixed carbon source in the form of a sugar (most often sucrose).

Table 1. Some of the elements important for plant nutrition and their physiological function. These elements have to supplied by the culture medium in order to support the growth of healthy cultures in vitro

Element	*Function*
Nitrogen	Component of proteins, nucleic acids and some coenzymes Element required in greatest amount
Potassium	Regulates osmotic potential, principal inorganic cation
Calcium	Cell wall synthesis, membrane function, cell signalling
Magnesium	Enzyme cofactor, component of chlorophyll
Phosphorus	Component of nucleic acids, energy transfer, component of intermediates in respiration and photosynthesis
Sulphur	Component of some amino acids (methionine, cysteine) and some cofactors
Chlorine	Required for photosynthesis

Iron	Electron transfer as a component of cytochromes
Manganese	Enzyme cofactor
Cobalt	Component of some vitamins
Copper	Enzyme cofactor, electron-transfer reactions
Zinc	Enzyme cofactor, chlorophyll biosynthesis
Molybdenum	Enzyme cofactor, component of nitrate reductase

One other vital component that must also be supplied is water, the principal biological solvent. Physical factors, such as temperature, pH, the gaseous environment, light (quality and duration) and osmotic pressure, also have to be maintained within acceptable limits.

Use of Culture Media

Culture media used for the in vitro cultivation of plant cells are composed of three basic components:

(1) essential elements, or mineral ions, supplied as a complex mixture of salts;

(2) an organic supplement supplying vitamins and/or amino acids; and

(3) a source of fixed carbon; usually supplied as the sugar sucrose.

For practical purposes, the essential elements are further divided into the following categories:

(1) macroelements (or macronutrients);

(2) microelements (or micronutrients); and

(3) an iron source.

Complete, plant cell culture medium is usually made by combining several different components, as outlined in Table 2.

Components of Media

Macroelements

As is implied by the name, the stock solution supplies those elements required in large amounts for plant growth and development. Nitrogen, phosphorus, potassium, magnesium, calcium and sulphur (and carbon, which is added separately) are usually regarded as macroelements. These elements usually comprise at least 0.1% of the dry weight of plants.

Table 2. Composition of a typical plant culture medium. The medium described here is that of Murashige and Skoog (MS)[a]

Essential element	Concentration in stock solution (mg/l)	Concentration in medium (mg/l)
Macroelements[b]		
NH_4NO_3	33,000	1,650
KNO_3	38000	1,900
$CaCl_2.2H_2O$	8800	440
$MgSO_4.7H_2O$	7400	370
KH_2PO_4	3400	170
Microelements[c]		
KI	166	0.83
H_3BO_3	1240	6.2
$MnSO_4.4H_2O$	4460	22.3
$ZnSO_4.7H_2O$	1720	8.6
$Na_2MoO_4.2H_2O$	50	0.25
$CuSO_4.5H_2O$	5	0.025
$CoCl_2.6H_2O$	5	0.025
Iron source[c]		
$FeSO_4.7H_2O$	5560	27.8
$Na_2EDTA.2H_2O$	7460	37.3
Organic supplement[c]		
Myoinositol	20000	100
Nicotinic acid	100	0.5
Pyridoxine-HCl	100	0.5
Thiamine-HCl	100	0.5
Glycine	400	2
Carbon source[d]		
Sucrose	Added as solid	30 000

[a] Many other commonly used plant culture media are similar in composition to MS medium and can be thought of as 'high-salt' media. MS is an extremely widely used medium and forms the basis for many other media formulations.

[b] 50 ml of stock solution used per litre of medium.

[c] 5 ml of stock solution used per litre of medium.

[d] Added as solid.

Nitrogen is most commonly supplied as a mixture of nitrate ions (from the KNO_3) and ammonium ions (from the NH_4NO_3). Theoretically, there is an advantage in supplying nitrogen in the form of ammonium ions, as nitrogen must be in the reduced form to be incorporated into macromolecules. Nitrate ions therefore need to be reduced before incorporation. However, at high concentrations, ammonium ions can be toxic to plant cell cultures and uptake of ammonium ions from the medium causes acidification of the medium. In order to use ammonium ions as the sole nitrogen source, the medium needs to be buffered.

High concentrations of ammonium ions can also cause culture problems by increasing the frequency of vitrification (the culture appears pale and 'glassy' and is usually unsuitable for further culture). Using a mixture of nitrate and ammonium ions has the advantage of weakly buffering the medium as the uptake of nitrate ions causes OH^- ions to be excreted. Phosphorus is usually supplied as the phosphate ion of ammonium, sodium or potassium salts. High concentrations of phosphate can lead to the precipitation of medium elements as insoluble phosphates.

Microelements

These elements are required in trace amounts for plant growth and development, and have many and diverse roles. Manganese, iodine, copper, cobalt, boron, molybdenum, iron and zinc usually comprise the microelements, although other elements such as nickel and aluminium are frequently found in some formulations. Iron is usually added as iron sulphate, although iron citrate can also be used. Ethylenediaminetetraacetic acid (EDTA) is usually used in conjunction with the iron sulphate. The EDTA complexes with the iron so as to allow the slow and continuous release of iron into the medium. Uncomplexed iron can precipitate out of the medium as ferric oxide.

Organic Supplements

Only two vitamins, thiamine (vitamin B_1) and myoinositol (considered a B vitamin) are considered essential for the culture of plant cells in vitro. However, other vitamins are often added to plant cell culture media for historical reasons. Amino acids are also commonly included in the organic supplement. The most frequently used is glycine (arginine, asparagine, aspartic acid, alanine, glutamic acid, glutamine and proline are also used), but in many cases its inclusion is not essential. Amino acids provide a source of reduced nitrogen and, like ammonium ions, uptake causes acidification of the medium. Casein hydrolysate can be used as a relatively cheap source of a mix of amino acids.

Carbon Source

Sucrose is cheap, easily available, readily assimilated and relatively stable and is therefore the most commonly used carbon source. Other carbohydrates (such as glucose, maltose, galactose and sorbitol) can also be used, and in specialised circumstances may prove superior to sucrose.

Gelling agents

Media for plant cell culture in vitro can be used in either liquid or 'solid' forms, depending on the type of culture being grown. For any culture types that require the plant cells or tissues to be grown on the surface of the medium, it must be solidified (more correctly termed 'gelled'). Agar, produced from seaweed, is the most common type of gelling agent, and is ideal for routine applications. However, because it is a natural product, the agar quality can vary from supplier to supplier and from batch to batch. Purified agar or agarose can be used, as can a variety of gellan gums.

Plant Growth Regulators

The essential point as far as plant cell culture is concerned is that, due to this plasticity and totipotency, specific media manipulations can be used to direct the development of plant cells in culture. Plant growth regulators are the critical media components in determining the developmental pathway of the plant cells. The plant growth regulators used most commonly are plant hormones or their synthetic analogues.

Classes of Plant Growth Regulators

There are five main classes of plant growth regulator used in plant cell culture, namely:

(1) auxins;

(2) cytokinins;

(3) gibberellins;

(4) abscisic acid;

(5) ethylene.

Auxins

Auxins promote both cell division and cell growth The most important naturally occurring auxin is IAA (indole-3-acetic acid), but its use in plant cell culture media is limited because it is unstable to both heat and light.

Occasionally, amino acid conjugates of IAA, which are more stable, are used to partially alleviate the problems associated with the use of IAA.

Table 3. Commonly used auxins, their abbreviation and chemical name

Abbreviation/name	Chemical name
2,4-D	2,4-dichlorophenoxyacetic acid
2,4,5-T	2,4,5-trichlorophenoxyacetic acid
Dicamba	2-methoxy-3,6-dichlorobenzoic acid
IAA	Indole-3-acetic acid
IBA	Indole-3-butyric acid
MCPA	2-methyl-4-chlorophenoxyacetic acid
NAA	1-naphthylacetic acid
NOA	2-naphthyloxyacetic acid
Picloram	4-amino-2,5,6-trichloropicolinic acid

It is more common, though, to use stable chemical analogues of IAA as a source of auxin in plant cell culture media. 2,4-Dichlorophenoxyacetic acid (2,4-D) is the most commonly used auxin and is extremely effective in most circumstances. Other auxins are available (Table 3), and some may be more effective or 'potent' than 2,4-D in some instances.

Cytokinins

Cytokinins promote cell division. Naturally occurring cytokinins are a large group of structurally related (they are purine derivatives) compounds. Of the naturally occurring cytokinins, two have some use in plant tissue culture media (Table 4).

Table 4. Commonly used cytokinins, their abbreviation and chemical name

Abbreviation/name	Chemical name
BAP[a]	6-benzylaminopurine
2iP (IPA)[b]	[N^6-(2-isopentyl)adenine]
Kinetin[a]	6-furfurylaminopurine
Thidiazuron[c]	1-phenyl-3-(1,2,3-thiadiazol-5-yl)urea
Zeatin[b]	4-hydroxy-3-methyl-trans-2-butenylaminopurine

[a] Synthetic analogues.

[b] Naturally occurring cytokinins.

[c] A substituted phenylurea-type cytokinin.

These are zeatin and 2iP (2-isopentyl adenine). Their use is not widespread as they are expensive (particularly zeatin) and relatively unstable. The synthetic analogues, kinetin and BAP (benzylaminopurine), are therefore used more frequently. Non-purine-based chemicals, such as substituted phenylureas, are also used as cytokinins in plant cell culture media. These substituted phenylureas can also substitute for auxin in some culture systems.

Gibberellins

There are numerous, naturally occurring, structurally related compounds termed 'gibberellins'. They are involved in regulating cell elongation, and are agronomically important in determining plant height and fruit-set. Only a few of the gibberellins are used in plant tissue culture media, GA_3 being the most common.

Abscisic acid

Abscisic acid (ABA) inhibits cell division. It is most commonly used in plant tissue culture to promote distinct developmental pathways such as somatic embryogenesis.

Ethylene

Ethylene is a gaseous, naturally occurring, plant growth regulator most commonly associated with controlling fruit ripening in climacteric fruits, and its use in plant tissue culture is not widespread. It does, though, present a particular problem for plant tissue culture. Some plant cell cultures produce ethylene, which, if it builds up sufficiently, can inhibit the growth and development of the culture. The type of culture vessel used and its means of closure affect the gaseous exchange between the culture vessel and the outside atmosphere and thus the levels of ethylene present in the culture.

Use of Plant Growth Regulators in Tissue Culture

Generalisations about plant growth regulators and their use in plant cell culture media have been developed from initial observations made in the 1950s. There is, however, some considerable difficulty in predicting the effects of plant growth regulators: this is because of the great differences in culture response between species, cultivars and even plants of the same cultivar grown under different conditions.

However, some principles do hold true and have become the paradigm on which most plant tissue culture regimes are based. Auxins and cytokinins are the most widely used plant growth regulators in plant tissue culture and are usually used together, the ratio of the auxin to the cytokinin determining the type of culture established or regenerated (Figure 1).

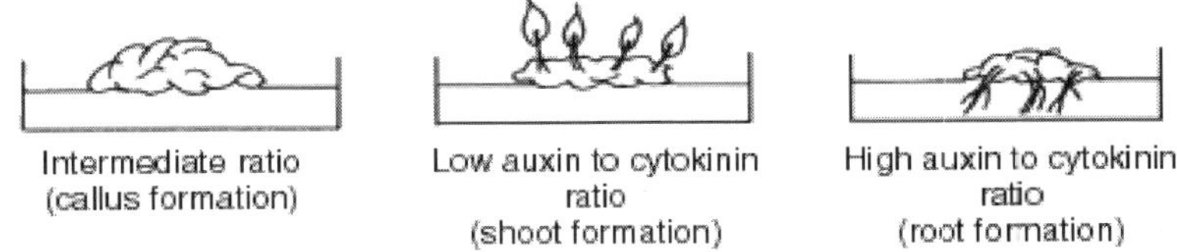

Fig. 1. The effect of different ratios of auxin to cytokinin on the growth and morphogenesis of callus

A high auxin to cytokinin ratio generally favours root formation, whereas a high cytokinin to auxin ratio favours shoot formation. An intermediate ratio favours callus production.

Types of Tissue Culture

Cultures are generally initiated from sterile pieces of a whole plant. These pieces are termed 'explants', and may consist of pieces of organs, such as leaves or roots, or may be specific cell types, such as pollen or endosperm. Many features of the explant are known to affect the efficiency of culture initiation. Generally, younger, more rapidly growing tissue (or tissue at an early stage of development) is most effective. Several different culture types most commonly used in plant transformation studies will now be examined in more detail.

Callus Culture

Explants, when cultured on the appropriate medium, usually with both an auxin and a cytokinin, can give rise to an unorganised, growing and dividing mass of cells. It is thought that any plant tissue can be used as an explant, if the correct conditions are found. In culture, this proliferation can be maintained more or less indefinitely, provided that the callus is subcultured on to fresh medium periodically. During callus formation there is some degree of dedifferentiation (i.e. the changes that occur during development and specialisation are, to some extent, reversed), both in morphology (callus is usually composed of unspecialised parenchyma cells) and metabolism.

One major consequence of this dedifferentiation is that most plant cultures lose the ability to photosynthesise. This has important consequences for the culture of callus tissue, as the metabolic profile will probably not match that of the donor plant. This necessitates the addition of other components-such as vitamins and, most importantly, a carbon source-to the culture medium, in addition to the usual mineral nutrients.

Callus culture is often performed in the dark (the lack of photosynthetic capability being no drawback) as light can encourage differentiation of the callus. During long-term culture, the culture may lose the requirement for auxin and/or cytokinin. This process, known as 'habituation', is common in callus cultures from some plant species (such as sugar beet). Callus cultures are extremely important in plant biotechnology. Manipulation of the auxin to cytokinin ratio in the medium can lead to the development of shoots, roots or somatic embryos from which whole plants can subsequently be produced. Callus cultures can also be used to initiate cell suspensions, which are used in a variety of ways in plant transformation studies.

Cell-suspension Culture

Callus cultures, broadly speaking, fall into one of two categories: compact or friable. In compact callus the cells are densely aggregated, whereas in friable callus the cells are only loosely associated with each other and the callus becomes soft and breaks apart easily. Friable callus provides the inoculum to form cell-suspension cultures. Explants from some plant species or particular cell types tend not to form friable callus, making cell-suspension initiation a difficult task. The friability of callus can sometimes be improved by manipulating the medium components or by repeated subculturing. The friability of the callus can also sometimes be improved by culturing it on 'semi-solid' medium (medium with a low concentration of gelling agent).

When friable callus is placed into a liquid medium (usually the same composition as the solid medium used for the callus culture) and then agitated, single cells and/or small clumps of cells are released into the medium. Under the correct conditions, these released cells continue to grow and divide, eventually producing a cell-suspension culture. A relatively large inoculum should be used when initiating cell suspensions so that the released cell numbers build up quickly. The inoculum should not be too large though, as toxic products released from damaged or stressed cells can build up to lethal levels. Large cell clumps can be removed during subculture of the cell suspension.

Cell suspensions can be maintained relatively simply as batch cultures in conical flasks. They are continually cultured by repeated subculturing into fresh medium. This results in dilution of the suspension and the initiation of another batch growth cycle. The degree of dilution during subculture should be determined empirically for each culture. Too great a degree of dilution will result in a greatly extended lag period or, in extreme cases, death of the transferred cells.

After subculture, the cells divide and the biomass of the culture increases in a characteristic fashion, until nutrients in the medium are exhausted and/or toxic by-products build up to inhibitory levels-this is called the 'stationary phase'. If cells are left in the stationary phase for too long, they will die and the culture will be lost. Therefore, cells should be transferred as they enter the stationary phase. It is therefore important that the batch growth-cycle parameters are determined for each cell-suspension culture.

Protoplast Culture

Protoplasts are plant cells with the cell wall removed. Protoplasts are most commonly isolated from either leaf mesophyll cells or cell suspensions, although other sources can be used to advantage. Two general approaches to removing the cell wall (a difficult task without damaging the protoplast) can be taken-mechanical or enzymatic isolation. Mechanical isolation, although possible, often results in low yields, poor quality and poor performance in culture due to substances released from damaged cells.

Enzymatic isolation is usually carried out in a simple salt solution with a high osmoticum, plus the cell wall degrading enzymes. It is usual to use a mix of both cellulase and pectinase enzymes, which must be of high quality and purity. Protoplasts are fragile and easily damaged, and therefore must be cultured carefully. Liquid medium is not agitated and a high osmotic potential is maintained, at least in the initial stages. The liquid medium must be shallow enough to allow aeration in the absence of agitation. Protoplasts can be plated out on to solid medium and callus produced. Whole plants can be regenerated by organogenesis or somatic embryogenesis from this callus. Protoplasts are ideal targets for transformation by a variety of means.

Root Culture

Root cultures can be established in vitro from explants of the root tip of either primary or lateral roots and can be cultured on fairly simple media. The growth of roots in vitro is potentially unlimited, as roots are indeterminate organs. Although the establishment of root cultures was one of the first achievements of modern plant tissue culture, they are not widely used in plant transformation studies.

Shoot Tip and Meristem Culture

The tips of shoots (which contain the shoot apical meristem) can be cultured in vitro, producing clumps of shoots from either axillary or adventitious buds. This method can be used for clonal propagation. Shoot meristem cultures are potential alternatives to the more commonly used methods for cereal regeneration as they are less genotype-dependent and more efficient (seedlings can be used as donor material).

Embryo Culture

Embryos can be used as explants to generate callus cultures or somatic embryos. Both immature and mature embryos can be used as explants. Immature, embryo-derived embryogenic callus is the most popular method of monocot plant regeneration.

Microspore Culture

Haploid tissue can be cultured in vitro by using pollen or anthers as an explant. Pollen contains the male gametophyte, which is termed the 'microspore'. Both callus and embryos can be produced from pollen. Two main approaches can be taken to produce in vitro cultures from haploid tissue. The first method depends on using the anther as the explant. Anthers (somatic tissue that surrounds and contains the pollen) can be cultured on solid medium (agar should not be used to solidify the medium as it contains inhibitory substances).

Pollen-derived embryos are subsequently produced via dehiscence of the mature anthers. The dehiscence of the anther depends both on its isolation at the correct stage and on the correct culture conditions. In some species, the reliance on natural dehiscence can be circumvented by cutting the wall of the anther, although this does, of course, take a considerable amount of time. Anthers can also be cultured in liquid medium, and pollen released from the anthers can be induced to form embryos, although the effi-ciency of plant regeneration is often very low. Immature pollen can also be extracted from developing anthers and cultured directly, although this is a very time-consuming process.

Both methods have advantages and disadvantages. Some beneficial effects to the culture are observed when anthers are used as the explant material. There is, however, the danger that some of the embryos produced from anther culture will originate from the somatic anther tissue rather than the haploid microspore cells. If isolated pollen is used there is no danger of mixed embryo formation, but the efficiency is low and the process is time-consuming. In microspore culture, the condition of the donor plant is of critical importance, as is the timing of isolation. Pretreatments, such as a cold treatment, are often found to increase the efficiency. These pretreatments can be applied before culture, or, in some species, after placing the anthers in culture.

Plant species can be divided into two groups, depending on whether they require the addition of plant growth regulators to the medium for pollen/anther culture; those that do also often require organic supplements, e.g. amino acids. Many of the cereals (rice, wheat, barley and maize) require medium supplemented with plant growth regulators for pollen/anther culture. Regeneration from microspore explants can be obtained by direct embryogenesis, or via a callus stage and subsequent embryogenesis.

Haploid tissue cultures can also be initiated from the female gametophyte (the ovule). In some cases, this is a more efficient method than using pollen or anthers. The ploidy of the plants obtained from haploid cultures may not be haploid. This can be a consequence of chromosome doubling during the culture period. Chromosome doubling (which often has to be induced by treatment with chemicals such as colchicine) may be an advantage, as in many cases haploid plants are not the desired outcome of regeneration from haploid tissues. Such plants are often referred to as 'di-haploids', because they contain two copies of the same haploid genome.

Plant Regeneration

Having looked at the main types of plant culture that can be established in vitro, we can now look at how whole plants can be regenerated from these cultures. In broad terms, two methods of plant regeneration are widely used in plant transformation studies, i.e. somatic embryogenesis and organogenesis.

Somatic Embryogenesis

In somatic (asexual) embryogenesis, embryo-like structures, which can develop into whole plants in a way analogous to zygotic embryos, are formed from somatic tissues (Figure 2). These somatic embryos can be produced either directly or indirectly. In direct somatic embryogenesis, the embryo is formed directly from a cell or small group of cells without the production of an intervening callus. Though common from some tissues (usually reproductive tissues such as the nucellus, styles or pollen), direct somatic embryogenesis is generally rare in comparison with indirect somatic embryogenesis.

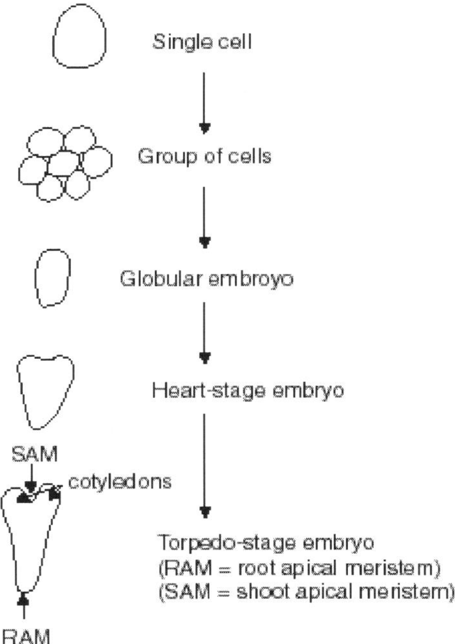

Fig. 2. A schematic representation of the sequential stages of somatic embryo development

In indirect somatic embryogenesis, callus is first produced from the explant. Embryos can then be produced from the callus tissue or from a cell suspension produced from that callus. Somatic embryogenesis from carrot is the classical example of indirect somatic embryogenesis. Somatic embryogenesis usually proceeds in two distinct stages. In the initial stage (embryo initiation), a high concentration of 2,4-D is used. In the second stage (embryo production) embryos are produced in a medium with no or very low levels of 2,4-D.

In many systems it has been found that somatic embryogenesis is improved by supplying a source of reduced nitrogen, such as specific amino acids or casein hydrolysate.

Somatic embryogenesis relies on plant regeneration through a process analogous to zygotic embryo germination. Organogenesis relies on the production of organs, either directly from an explant or from a callus culture. There are three methods of plant regeneration via organogenesis. The first two methods depend on adventitious organs arising either from a callus culture or directly from an explant (Figure 3). Alternatively, axillary bud formation and growth can also be used to regenerate whole plants from some types of tissue culture.

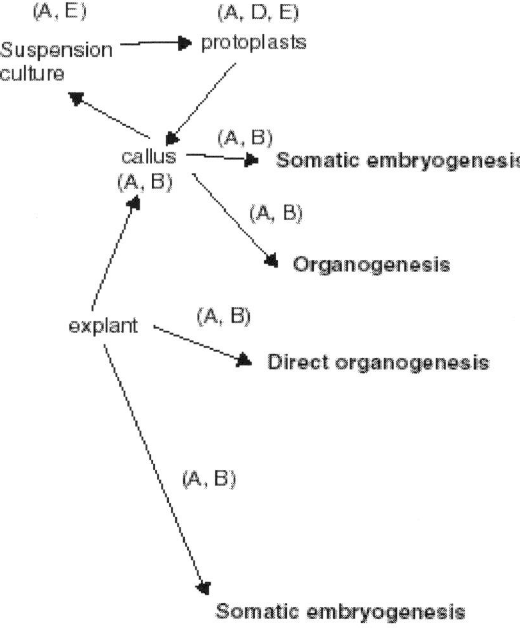

Fig. 3. A simplified scheme for the integration of plant tissue culture into plant transformation protocols

Organogenesis relies on the inherent plasticity of plant tissues, and is regulated by altering the components of the medium. In particular, it is the auxin to cytokinin ratio of the medium that determines which developmental pathway the regenerating tissue will take. It is usual to induce shoot formation by increasing the cytokinin to auxin ratio of the culture medium. These shoots can then be rooted relatively simply.

Integration of Plant Tissue Culture

Various methods of plant regeneration are available to the plant biotechnologist. Some plant species may be amenable to regeneration by a variety of methods, but some may only be regenerated by one method. The various methods that can be used to transform plants will be considered, but it is worthwhile briefly considering the interaction of plant regeneration methodology and transformation methodology here (Fig. 3). Not all plant tissue is suited to every plant transformation method, and not all plant species can be regenerated by every method. There is therefore a need to find both a suitable plant tissue culture/regeneration regime and a compatible plant transformation methodology.

Applications of Tissue Culture

Plant tissue culture, the growth of plant cells outside an intact plant, is has direct commercial applications that include:

— Micropropagation used in forestry, and to conserve rare or endangered plant species.

— A plant breeder using tissue culture to screen cells rather than plants for advantageous characters, e.g. herbicide resistance/tolerance.

— Large-scale growth of plant cells in liquid culture inside bioreactors as a source of secondary products, like recombinant proteins used as biopharmaceuticals.

— Crossing distantly related species by protoplast fusion and regeneration of the novel hybrid.

— Cross-pollinating distantly related species and tissue culturing the resulting embryo that would otherwise normally die.

— Production of doubled monoploid (dihaploid) plants from haploid cultures to get homozygous lines more rapidly in breeding programmes, usually by treatment with colchicine that causes doubling of the chromosome number.

— As a tissue for transformation, followed by either short-term testing of genetic constructs or regeneration of transgenic plants.

- Employing certain techniques such as meristem tip culture that can be used to produce clean plant material from virused stock, such as potatoes and many species of soft fruit.
- Micropropagation using meristem and shoot culture to produce large numbers of identical individuals
- Screening programmes of cells, rather than plants for advantageous characters
- Large-scale growth of plant cells in liquid culture as a source of secondary products
- Crossing distantly related species by protoplast fusion and regeneration of the novel hybrid
- As a tissue for transformation, followed by either short-term testing of genetic constructs or regeneration of transgenic plants
- Removal of viruses by propagation from meristematic tissues.

Production of Haploids

Haploids can produced *in vitro* thus solving some problems in genetic studies since gene actions are readily manifested due to a single allelic dose present in chromosomes of an entire genome. This provides them suitable material for induction of mutations and various gene manipulations. Through rapid achievement of homozygous traits in double-haploids, pollen-derived haploid plants have been used in breeding and improvement of crop species.

Release New Varieties through F1 Double-haploid System

One cycle of meiotic recombination is usually insufficient for the improvement of agronomic quantitative traits. This is because the linkage between polygenes will not release all potential variations available in a cross. Combining anther-culture with sexual hybridisation among different genotypes of anther- derived plants has helped to overcome this problem. The anthers of the hybrid (F1) progeny are excellent breeding material for raising pollen-derived homozygous plants in which complementary parental characteristics are combined in one generation.

Studies related to inheritance of qualitative traits show that in breeding programmes the double-haploids derived from pollen cultures expressed genetic variability to an extent that new varieties have been synthesised in respect of barley, brassica, rice, maize, rye, potato, pepper and asparagus. By inducting and haploid then by chromosome doubling, exclusively male plants in dioecious species can be obtained.

Selection of Mutants Resistant to Disease

Screening mutants with resistance to disease is of prime importance in crop improvement. Haploids can be used in rapid selection of mutant that disease-resistant because they provide a relatively easier system for the induction of mutations. Tobacco mutants resistant to black shank disease and wheat lines resistant to scab are some examples of using anther culture technique in mutant selection successfully.

Developing Asexual Lines

A pollen-derived rubber tree that was taller by six metres multiplied by asexual propagation to raise several clones. Particularly interesting in the poplar tree species is that haploid seedlings selected for desired genotypes are naturally doubled as diploids after 7-8 years, this coinciding with their flowering and enabling them to be both asexually and sexually propagated. Pollen-derived haploid plantlets have also been produced in other perennial woody species too.

Transfer of Desired Alien Genes

The chromosomal instability in haploids attracts alien chromosomes or genes during wider crossing programmes. In rice, the breeding of high-yielding varieties for resistance to blast is conventionally achieved through back-cross which is a time-consuming process, taking probably more than twelve years to develop a cultivar with desired resistance. Hybridisation and anther culture can ensure that a standard pollen-plant breeding procedure can be developed in rice to introduce genes for high yield and resistance to blast in a short span of two years.

Somacional Variations

Genetic variations occur in undifferentiated cells, isolated protoplasts, calli, tissues and morphological traits of regenerated plants mostly due to changes in the chromosome number and structure. Cytological heterogeneity in cultures arises due to: The expression of chromosomal mosaicism or genetic disorders in cells of the initial explants, and new irregularities brought about by culture conditions.

Cell or tissue cultures undergo frequent genetic changes (polyploidy, aneuploidy, chromosomal breakage, deletion, translocation, gene amplifications and mutations) which are also expressed at biochemical or molecular levels. Plant cell and tissue cultures, therefore, provide increased genetic variability relatively quick and with a simple technology. The genetic variability in cultures expresses in the form of variant traits in regenerated plants which are then transmitted to the progeny through sexual or vegetative propagation. This genetic variability is called somaclonal variation or gametoclonal variation. Plants regenerated from tissue and cell cultures show heritable variation for both qualitative and quantitative traits. Somaclonal variation has been described in sugarcane, potato, tomato etc.

Isolation of Somaclonal Variants

It is far easier to isolation mutants for several traits from cell cultures than from whole plant populations since a large number of cells, can be easily and effectively screened for mutant traits which would otherwise be very difficult, ordinarily impossible. Mutants can be selectively chosen for disease resistance, improvement of nutritional quality, adaptation of plants to stress conditions, e.g., saline soils, low temperature, toxic metals (e.g., aluminium), resistance to herbicides and to enhanced the biosynthesis of plant products used for medicinal or industrial purposes. Isolation of somaclonal variants can be done by screening and cell selection.

i) *Screening.* This technique which involves the observation of a large number of cells or regenerated plants for the detection of variant individuals is the only feasible approach for the isolation of mutants for yield and yield traits. Screening has been profitable for the isolation of cell clones that produce higher quantities of certain biochemicals and for computer-based automated cell sorting devices.

ii) *Cell selection.* In this approach, a suitable selection pressure permits the preferential survival/growth of variant cells only, e.g., selection of cells resistant to various toxins, herbicides, high salt concentration etc. When the selection pressure allows only the mutant cells to survive or divide, it is called positive selection and where the wild type cells divide normally and therefore are killed by a counter selection agent it is called negative, selection. The mutant cells, unable to divide escape the counter selection agent are subsequently rescued by removal of the counter selection agent. The negative selection approach is utilised for the isolation of auxotrophic mutants.

The positive selection approach may be subdivided into four categories: (i) direct selection, (ii) rescue method, (iii) stepwise selection and (iv) double selection.

In direct selection, the selection-pressure-resistant cells survive and divide to form colonies while the wild type cells are killed by the selection agent. This common selection method is used for the isolation of cells resistant to toxins (produced by pathogens), herbicides, elevated salt concentration, antibiotics, amino acid analogues etc.

As in the direct selection in the rescue method too, the wild type cells are killed by the selection agent, and the variant cells remain alive but, usually, they do not divide due to the unfavourable environment. The selection agent is then eliminated to save the variant cells. This approach has been used to recover low temperature and aluminium-resistant variant cells.

In the selection pressure, e.g., salt concentration, which may be gradually increased from a relatively low level to the cytotoxic level, the resistant clones isolated at each stage are subjected to the higher selection pressure. Such a selection approach, called stepwise selection, may often favour gene amplification or mutations in the organelle DNA.

In certain cases, it may be viable to select for survival and/or growth and also some other feature reflecting resistance to the selection pressure; this is called double selection. An example of double selection is antibiotic streptomycin, which inhibits chlorophyll development in cultured cells. This approach has been used for the selection of cells resistant to the herbicide amitrole, tobacco mosaic virus (TMV) and aluminium.

Characterisation of Variants

Somaclonal variants show instability when they are isolated through cell selection. The frequency of stable variants may range from eight per cent to 62 per cent, perhaps depending on the species and the selection agent. During further screening or selection many selected clones fail to exhibit their resistance and are susceptible, misclassified as resistant.

Several clones, called unstable variants, lose their resistance to the selection agent after a period of growth in the absence of selection pressure. They may result from changes in gene expression and from gene amplification.

Some variant phenotypes, while quite stable during the cell culture phase, disappear when plants are regenerated from the variant cultures, or when the regenerated plants reproduce sexually, in case they are expressed in the regenerated plants. Such epigenetic changes are attributed to stable changes in gene expression e.g., hormone habituation of cell cultures. The remaining variants that stably express the variant phenotypes during the cell culture and the regenerated plant phases, and exhibit the transmission of these phenotypes through the sexual reproduction cycle are called mutants. Only such a category of variants would find an application in crop improvement and may represent true gene mutations or some other types of changes.

Molecular Basis of Somaclonal Variation

Somaclonal variation may arise due to changes in chromosome number and/or structure, gene mutation, plasmagene mutation, alteration in gene expression, gene amplification, mitotic crossing over, transposable element activation, and rearrangements in cytoplasmic genes.

Most mutants isolated from cell cultures involving single gene mutations, the mutant allele is either dominant or recessive. Gene amplification in some variants, are recovered though stepwise selection of plant cells *in vitro*. Sometimes, deamplification may also occur in somaclonal variants, e.g., for rRNA genes. Also, some previously silent mutator genes may get activated during tissue culture e.g., in maize. Clearly, transposable elements can be activated during *in vitro* culture. The breakage and fusion of chromosomes, which occur during culture, may activate some controlling elements. Mitotic crossing-over could also be responsible for some of the genetic variations detected in regenerated plants leading to the recovery of homozygous recessive single gene mutations in plants regenerated from cells *in vitro*.

Somaclonal Variations and Induced Mutations

In general, cell cultures are not given mutagenic treatments for the recovery of somaclonal variants. But where mutagenic treatments were used, usually an increase in the frequency of somaclonal variants was observed. While in some cases, mutagenesis is necessary for the recovery of the specific variant being isolated, it should be avoided as far as possible in view of the undesirable features associated with such treatments. Somaclonal variations are preferable to induced mutations for several reasons:

i) Chimareism is a major problem in induced mutations but not in somaclonal variations.

ii) Induced mutations are often linked with undesirable features like sterility etc.

iii) 'New' alleles and even 'new' mutations have been recovered through somaclonal mutations.

iv) The frequency of useful mutations is very high in the case of somaclonal variations.

v) A highly effective selection can be applied *in vitro* for several economically important characters which is virtually impossible in the case of mutation breeding,

vi) An astronomically large number of individuals can be effectively screened *in vitro*, etc. But, it may be stressed that the somaclonal variation applies to only those species where whole plant can be regenerated from cultured cells, while mutation breeding applies to all the species, and that the former is dependent on quite sophisticated facilities for tissue culture and of greenhouse.

Gametoclonal Variations

Haploid plants regenerated from callus cultures show gametoclonal variation like in rice. Such a variation may be subjected to selection at the haploid level, and the chromosome number of the selected plants may be doubled to obtain homozygous plants.

Gametoclonal variation appears to be more desirable than somaclonal variation for the following reasons:

i) The mutant characteristic is expressed in the Ro plants.

ii) Cells with detrimental mutations may regenerate much less frequently in the case of haploid than in diploid cells. However, regeneration of haploid plants is feasible in fewer plant species than regeneration from somatic cells.

Somatic-derived somaclones and gametic-derived gametoclones can be distinguished in three ways:

i) Both dominant and recessive mutant genes induced by gametoclonal variation will express directly in haploid plants regenerated from microspores of diploid anthers.

ii) Recombinants recovered in gametoclones would be the result of meiotic crossing-over.

iii) The gametoclone can be used only after having stabilised by doubling its chromosome number.

The value of gametoclonal variation in crop improvement is evident from the development of double-haploid lines by anther culture of F1 hybrid plants of wheat and rice. Based on certain observations, it is suggested that factors causing meiotic recombination and mutation prior to initiating anther culture give rise to gametoclonal variation. Mutations may, however, occur even after or by the doubling process. Further, the genetic changes associated with the use of microspores for regeneration equally contribute to gametoclonal variation.

Variation in chromosome number of gametes or gametophytic tissue has a vital role in gametoclonal variation as evident from a range of aneuploids and mixoploids recovered from anther cultures of wheat, maize, and sexual hybrid of wheat and triticale. Also, variants uncovered after androgenesis are never recovered from somatic protoplast cultures. Further, it is observed that gene amplification contributes to reduction in the yield of gametoclonal variant plants. Thus, the mutation spectrum obtained by gametoclonal variation may differ from that obtained by somaclonal variation.

Applications in Plant Breeding

Somaclonal variation and gametoclonal variation are useful sources of introducing genetic variations valuable to the plant breeders.

Breeding of autotetraploid (4x) crops like potato, alfalfa etc. presents problems due to the complex segregation patterns at the autotetraploid level. Whereas breeding at the haploid level of crops like potato is much easier due to much simpler segregation patterns and smaller breeding populations involved. Haploids of tetraploid (4x) species are called dihaploid since they have two similar or dissimilar genomes. Potato dihaploids (2x) are fully fertile, and allow hybridisation and selection at this level. The selected dihaploid clones are given colchicine treatment to obtain tetraploid clones which are used as varieties. The process involving extraction of dihaploids from tetraploid species, breeding at the dihaploid level and then chromosome doubling of selected dihaploid lines to obtain tetraploid varieties is called analytical breeding which is becoming increasingly popular in crops like potato. Use of dihaploids in breeding of crops like potato offers the following advantages:

i) much simpler segregation ratios and, as a result, smaller segregating populations,

ii) a greater efficiency of selection, and
iii) much easier hybridisation with diploid wild species which offer many genes of great value.

There are several effective schemes for analytical breeding, including a scheme for extraction of monoploids (monoploids have a single genome) and then protoplast fusion of four selected monoploids to obtain autotetraploid varieties in crops like potato.

The following are the advantages:

i) Production of homozygous DH lines using haploids saves at least four years in comparison to selfing/close inbreeding.
ii) Selection among DH lines is at least 6-8 times as efficient as that among segregating populations thus reducing the size of breeding populations. In addition, the confusing effects of heterozygosity are absent in DH lines.
iii) Haploids may be useful for the isolation of desirable mutants whose chromosome number may be doubled to obtain homozygous mutant lines in a single generation.
iv) Pollen embryos are generally highly regenerative. Therefore they can be used for gene transfers mediated by agrobacterium or by a technique like particle gun.

But analytical breeding also has its limitations, such as:

i) In many crops, the application of this technique is not yet available, hence not possible for haploid production.
ii) In many other crops, since large numbers of haploid plants are not easily obtained a wider application of pollen culture in crop improvement depends primarily upon the development of techniques to produce more and quicker methods.
iii) High cost of obtaining haploids and doubled haploids is still a major problem.
iv) Sometimes deleterious mutations may be induced during the *in vitro* phase.
v) A non-random recovery of haploids may reduce the spectrum of variability recovered.
vi) A sophisticated tissue culture laboratory and a dependable greenhouse are essential for success.

vii) To carry out the various operations specialised skill is required.

viii) Plants other than haploids are often recovered from anther cultures/ interspecific hybrids; this necessitates cytological analyses.

ix) Occurrence of grametoclonal variation may limit the usefulness of pollen embryos for genetic transformation/ gene transfer.

x) The problem of high frequency of albinos produced in anther cultures of monocots, especially cereals needs to be resolved for a successful exploitation of haploids in breeding of such crops.

Germplasm Conservation

With population explosion, pressure on the forest and land resources there has been a decline in the population of medicinal and economically important plant species, some of which are on the verge of vanishing. Such 'threatened' species are trying to be protected and preserved for future breeding programme.

In India, the Botanical Survey of India has brought out three volumes highlighting the threatened plants of the Indian subcontinent. Several other organisations, are engaged in the protection and preservation of such plants. Preserving the genetic material can be done in two ways:

i) *In situ preservation:* This aims at the preservation of the germplasm in their natural environment by establishing biosphere reserves, national parks, gene sanctuaries etc. This type of preservation there is the risk of declination of the preserved species due to environmental hazards.

ii) *Ex situ preservation:* Being the chief mode of preservation of germplasm, it provided a suitable condition in the gene bank to preserve the genetic materials in the form of seed or *in vitro* cultures. But this requires considerable knowledge of genetic structure as well as elements influencing them.

Conserving plant germplasm in the seed-propagated plants has certain limitations such as: Loss of seed viability with passage of time, seed destruction due to seed born pathogens, pest etc., and is confinement only to seed propagating plants.

In contrast the vegetatively propagated plants are preserved *in vitro* as shoots, meristems, embryos etc. The advantages of *in vitro* this over in situ preservation are that large amount of material can be preserved in small area as; it overcomes the destruction due to environmental hazards; and it provides large amounts of plant material for culturing.

Cryopreservation

In cryopreservation metabolic functions of biological material are reduced and subsequently arrested by imposition of ultra-low temperature. At the temperature of liquid nitrogen (–196° C) almost all metabolic activities of cells cease and the sample is then preserved in such a state for extended periods. However, only certain biological materials, in their natural state can be frozen to sub-freezing temperatures without adversely affecting the cell viability. Knowledge about the chemicals having cryopreservative properties such as glycerol and dimethyl sulfoxide has facilitated the development of effective cryopreservation technique.

Need for Cryopreservation

The main purpose and aim of developing cryopreservation technology is to preserve valuable genetic resources, especially of vegetatively propagated and also of those species which have short lived seeds. Therefore, it is essential to have efficient regeneration protocols through tissue culture of the species. Regeneration methods using apical meristem are advantageous as these are simple and there are no chances of genetic variation. It is important that tissue culture methods do not create genetic variability.

In contrast in cases where genetic variability has been induced in the cultures, this variability needs preservation for use in the future. Therefore, cryopreservation technique is equally good for preserving the genetic resources of existing genotypes and also of new variants.

Procedure of Cryoprotection and Pretreatment

Cryoprotection in some form is necessary for cryopreservation of plant materials unless they are naturally dehydrated, as in the case of dormant vegetative buds in the winter, or artificially cold acclimated. Several chemicals, such as dimethyl sulphoxide (DMSO), glycerol, various sugars and sugar alcohols, protect living cells against damage during freezing and thawing. The colligative properties of the cryoprotectants lower the temperature at which freezing first occurs and can alter the crystal habit of ice when it separates, thus minimising the harmful action of electrolyte concentration resulting from conversion of water into ice. Cryoprotectants having high solubility in aqueous phase and low toxicity to the cells are categorised as: Permeating, e.g. DMSO, methanol, glycerol; and non-permeating, e.g., sugars, sugar alcohols, high molecular weight polymers as dextran, polyvinyl pyrrolidone, hydroxy ethyl starch.

The cells require different pretreatment periods with different compounds for proper cryoprotection; e.g., DMSO enters more rapidly than glycerol, in permeating types, and therefore requires shorter period for treatment. Most of the cryoprotectants exhibit varying degrees of cytotoxicity at higher concentrations, with DMSO at 5 to 10 per cent and glycerol at 10 to 20 per cent. Sometimes, a mixture of cryoprotectants improves the efficacy, e.g., addition of osmotically active compounds in the culture medium such as mannitol, sorbitol, sucrose and proline increase the freezing resistance of the cells acting by their dehydration effect. Mannitol beneficial in reducing the mean cell volume of cells of *Acer pseudoplatanus* and *Capsicum annuum* and increase their post-freezing survival. Therefore, all such compounds also possess cryoprotective properties besides that they are osmotically active.

Freezing methods

Various plant materials can be cryopreserved by slow freezing, rapid freezing and droplet-freezing.

i) *Slow freezing*: When plant cells are cooled progressively, ice crystal formation begins extra-cellularly, presuming that plasma membrane acts as a barrier that prevents the ice crystal formation in the cytoplasm. In the absence of ice crystals, cytoplasm remains super cooled. On reducing temperature further, the concentration of extra-cellular liquid increases, as more water is converted to ice. The vapour pressure of the frozen solution being lower than the same concentration of the super cooled liquid, its of slowly cooled cells reach equilibrium with external ice by efflux of water thus, preventing the intra-cellular ice formation and consequently preventing freezing injury. It is believed that slow freezing increases the concentration of cytoplasm and increased dehydration increases the survival of cells. The development of efficient slow freezing method depends upon several factors like: Cooling rates; pretreatment and cryoprotection; type and physiological state of the material; and the temperature prior to immersion in liquid nitrogen. The most commonly used methods generally involves regulated slow cooling at a constant rate of 0.5 to 2° C/min. to terminal temperature between –3° C to -40° C followed by storage in liquid nitrogen.

ii) *Rapid freezing:* Rapid freezing, unsuitable for the cryopreservation of cell cultures, is employed to cryopreserve shoot tips of carnation, potato, strawberry and several others, and somatic emboyos of oil palm. Rapid freezing is done by direct immersion of the cryoprotectant-treated specimens in liquid nitrogen. At such high cooling rates, the intracellular fluids do not have sufficient time to equilibrate with the external ice with the possibility of intracellular ice formation, which is considered to be harmful for cells.

iii) *Droplet freezing:* In this method, the cryoprotectant treated meristems are dispensed in droplets of 2-3 ml on an aluminium foil in a petriplate being frozen by slow cooling (0.5° C/min.) to a subzero temperature between –20 to –40° C prior to immersion in liquid nitrogen.

Material can be kept stored in liquid nitrogen (–196°C) or in its vapour (–150°C), and rapidly thawed is to avoid the damaging ice recrystallisation which may occur during slow warming. In general, thawing is carried out by removing the sample by liquid nitrogen storage and transferring in a water bath (34–40°C) for about 1–2 minutes or until material is warmed up.

The most reliable and accurate estimate of viability is regrowth of cryopreserved specimens. The other technique used for vitality test of the cells such as fluoresce in diacetate stain, triphenyl tetrazolium chloride test etc. can give a quicker method of testing cell viability. The further growth and regeneration of the cryopreserved cells after thawing will be like normal cells which should be carefully handled, plated and while subculturing of such cells.

Production of Secondary Metabolities

Plant cell culture can be a potential means of producing useful plant products thereby doing away with conventional agriculture with all its attendant problems and variables like: environmental factors (drought, floods, etc.), disease, political and labour instabilities in the producing countries, uncontrollable variations in the crop quality, inability of authorities to prevent crop adulteration, losses in storage and handling, etc. Thus, the production of useful and valuable secondary metabolites in large bioreactors located in the consuming country leads to controlled production according to demand and a reduced time requirement.

Theoretically, that such large scale suspension cultures will be suitable for industrial production of useful plant chemicals such as pharmaceuticals and food additives, similar to that of microbial fermentation. Yet due to some significant differences between microbial and plant cell cultures consideration must be given attempting to apply plant cell cultures to the available technology.

Table 5 shows a comparison of some of the characteristics of plant and microbial cultures of relevance to fermentation. This table serves to demonstrate some of the problems that can be encountered with plant cell cultures. The sensitivity to shear is due both to the large size of the cells and to the relatively inflexible cellulose cell wall. Thus, with normal blade impellers the cells may twist which will inhibit mitoses and, for this reason, air- lift fermentors are recommended by some researchers. The large size of the plant cell contributes to its comparatively high doubling time, which thus prolongs the time required for a successful fermentation run.

Table 5. Characteristic of Microbial and Plant Cell Relevant to Fermentation

Characteristics	*Microorganism*	*Plant Cell*
Size	2 u	>10 u
Shear stress	Insensitive	Sensitive
Water content	75%	>90%
Duplication time	<1 hour	days
Aeration	1-2 vvm	0.3 vvm
Fermentation time	Days	Weeks
Product accumulation	Medium	Vacuole
Production phase	Uncoupled	Often growth-linked
Mutation	Possible	Requires haploids
Medium cost ($) (MS medium)	8-9/m	65-70/m

Since product secretion is uncommon in a vocuole, the high metabolite yields seen in microorganisms that secrete product cannot be expected. Membrane permeabilisation of plant cells may serve to relieve the constraints of product inhibition by facilitation of leakage into the extracellular medium, and by permitting recycling of the biomass it would help reduce production costs.

The low aeration requirement for plant cells is an advantage over microbial cultures in general. In addition, the high cost of running a fermentation vessel over several weeks should be considered, although media costs are much less than those of animal cell cultures.

In addition to the above problems concerned with fermentation technology, at the biochemical level, the two major problems concern poor expression of products and instability of cell lines.

Compared with intact plant cells cultured plant cells often produce reduced quantities and different profiles of secondary metabolites and these quantitative and qualitative features may change with time. The poor product expression may often be due to a lack of differentiation in cultures and in some cases of cultures there is over-production of metabolites compared with the whole plant.

Many cultured cells producing metabolites not observed in the plant, e.g., Lithospermum erythrorhison cultures synthesizing rosmarinic acid. It is clear that selecting original plant material having high yields of the desired phytochemical may be important in establishing high-yielding cultures. Furthermore, it is important to screen for high producing lines over and over again as also the nutritional.

Materials

any part obtained from any plant species can be employed to induce callus tissue, however the successful production of callus depends upon plant species and their qualities. Dicotyledons are rather amenable for callus tissue induction, as compared to monocotyledons; the callus of woody plants generally grow slowly. Stems, leaves, roots, flowers, seeds and any other parts of plants are used, but younger and fresh explants are preferable as explant materials. Explants obtained must be sterilised using ethanol, sodium hypochlorite and/or other chemicals to remove all microorganisms from the materials and a typical sterilisation procedure will be described later as an example.

Media

Inorganic Salts

Many media are designed induce a callus from an explant and to cultivate the callus and cells in suspension, agar or its substitutes into the media to prepare solid medium for callus induction. A most commonly used media for plant tissue cultures is one that has very high concentration of nitrate, potassium and ammonia. In comparison, the levels of inorganic nutrients in the B5 medium are lower. Many other media have been developed and modified and nutrient however, it is not always necessary to test many kinds of basal media when a callus is induced, rather using only one or two kinds of basal media in combination of different kinds and concentrations of phytohormones. The most suitable medium composition should be optimised afterwards in order to obtain higher level of products and higher growth rate.

Carbon Sources

By adding suitable carbon sources such as sucrose or glucose, fructose, maltose and other sugars also support the growth of various plant cells is enhanced and the efficient production process of useful metabolites is established depending on plant species and products; therefore it is necessary to optimise the medium compositions including carbon sources in each case. From an economically also, the use of more inexpensive carbon sources is appropriate in industry and crude sugars such as molasses have been examined.

Vitamins

The basal media include myo-inositol, nicotinic acid, pyridoxine HCl and thiamine HCl thiamine being essential one for many plant cells, with other vitamins stimulating the growth of the cells in some cases.

Phytohormones

Phytohormones or growth regulators are required to induce callus tissues and to promote the growth of many cell lines. The concentration of auxins in the medium is generally between 0.1 and 50 μM. Since each plant species requires different kinds and levels of phytohormones for callus induction, its growth and metabolites production, it is important to select the most appropriate growth regulators and to determine their optimal concentrations. Gibberellic acid is also added to the medium if necessary.

Organic Supplements

In order to stimulate the growth of the cells, organic supplements like casamino acid, peptone, yeast extracts, malt extracts and coconut milk are sometimes added to the medium.

Large Scale Cultures

Commercial applications of tissue culture requires scaling up of culture systems; which can be achieved by (i) increasing the number or (ii) size of culture vessels, or (iii) a combination of both, depending chiefly on the physical condition of medium, i.e., agar or liquid, used for the cultures.

Agar Cultures

The most commonly used medium for experimental and some commercial activities like micropropagation, is agar-geiled. Generally during experimentation, scaling up can be achieved by increasing the number of culture vessels. Test tube baskets or stands containing culture tubes which are placed on culture racks. Half litre bottles are the most commonly used culture vessels in commercial micropropagation activities. While scaling up agar cultures each vessel has to be handled individually during subculture.

Suspension Cultures

Use of liquid medium in suspension cultures allows easy and extensive scaling up by employing bioreactors, although a limited scaling up can be achieved by using larger culture flasks.

A bioreactor is a culture vessel, generally of a large volume, e.g., 1L to over 1000L which has provisions for:

i) aeration,

ii) stirring to achieve medium and cell mixing,

iii) contamination control and

iv) replacement of used medium and/or used medium plus cells.

The bioreactors used for culture of plant cells may be of following four types: (a) batch bioreactors, (b) continuous bioreactors, (c) multistage bioreactors and (d) immobilised cell bioreactors; all bioreactors, except the last one are commonly called stirred tank reactors since the vessel has a device for stirring.

Batch bioreactors. In batch bioreactors, the medium and inoculum are loaded at the outset. The cells are allowed to grow and the entire cell mass is harvested at the end of incubation period. The characteristic features of such bioreactor systems are: continuous depletion of medium, accumulation of cellular wastes, alterations in growth rate, and continuous change in the composition of cells. The spin filter bioreactor can be used as a batch bioreactor by closing the inlet for medium and the outlets for medium/ medium plus cells.

The bioreactor has ports for (1) addition of fresh medium, (2) air space, (3) removal of only spent medium (minus cells; they are separated by the spin-filter) through the hollow central shaft, (4) entry of air to the sparger, and (5) removal of spent medium plus cells. When all the 5 ports are in operation, the bioreactor is of continuous flow type. But when only ports (2) and (4) (air outlet and inlet, respectively) are open with remaining ports of (1), (3) and (5) closed, the bioreactor becomes a batch type.

Continuous bioreactors. In continuous bioreactors, during the entire incubation period there is continuous inflow of fresh medium and outflow of used medium. A spin-filter bioreactor is a good example of continuous flow bioreactor.

The continuous bioreactor provides a highly versatile system for control on medium change rate and on cell density, being made possible due to the two routes for medium removal while only one of them allows the removal of cells. Using a continuous flow bioreactor are grown cells at a specified cell density in an active growth phase. Such cultures may either provide inocula for further culture or may serve as a continuous source of biomass yields.

Multistage bioreactors. Using two or more bioreactors in a specified sequence in such culture systems each carries out a specific step of the total production process.

A perfusion type bioreactor allows regular circulation or change of medium by regularly removing the used medium only through the spin-filter and adding fresh medium through the medium port. It also allows change of one type of medium by another as per the need of the culture system. The simplest situation would involve two bioreactors to produce a biochemical like shikonin, the first of the two bioreactor provides conditions for rapid cell proliferation and favours biomass production, while the other one has conditions conducive for shikonin biosysnthesis and accumulation. The cell biomass, is collected from the first stage bioreactor, is used as inoculum for the second stage reactor. Similarly, for large scale somatic embryo (SE) production, the first reactor may be in continuous mode, while the second may be of batch type. The continuous first stage reactor provides conducive conditions for rapid proliferation of embryogenic cells. The cell mass from this bioreactor serves as a continuous source of inoculum for the second stage batch type bioreactor which has suitable conditions for embryo development and maturation.

The use of continuous first stage bioreactor (i) avoids the time, labour and cost needed for cleaning etc. of a batch reactor between two runs, (ii) eliminates the lag phase of batch cultures, and (iii) provides a more homogeneous and actively growing cell population. However, due to their long run periods, continuous cultures have a greater contamination risk than batch bioreactors.

Immobilised cell bioreactors: These bioreactors are based on cells encased either in gels, such as, agarose, agar, chitosan, gelatine, gellan, polyacrylamide and calcium alginate to produce beads, or in a membrane or metal screen compartment or cylinder. The cell-containing membrane/ screen cylinder is kept in a chamber through which the medium is circulated from a recycle chamber. The medium movement may be so adjusted as to flow across the screen compartment rather than parallel to it. The technology is being refined for commercialisation. Fresh medium is regularly added to and equivalent volume of used medium is withdrawn from the recycling chamber to maintain its nutrient status.

The technique cell immobilisation changing the physiology of cells as compared to that of cells in suspension is useful where the biochemical of interest is excreted by the cells into the medium. Product excretion may also be effected by immobilisation itself, or by certain treatments like altered pH, use of DMSO (dimethyl sulfoxide) as a permealising agent, changed ionic strength of medium, an elicitor etc. Immobilised cell reactors have no risk of cell wash out, have low contamination risk, protection of cells from liquid shear, better control on cell aggregate size, separation of growth phase (in a batch/continuous bioreactor) from production stage (in an immobilised cell bioreactor), cellular wastes regularly removed from the system, and cultures at high cell densities.

Plant Cell Culture for Production of Chemicals

A large variety of biochemicals which are metabolites of both primary and secondary metabolism are derived from plants. But, secondary metabolites have impressive biological activities like antimicrobial, antibiotic, insecticidal, molluscidal, hormonal properties, and valuable pharmacological and pharmaceutical activities, and many are used as flavours, fragrances, colours, etc. Secondary metabolites include a variety of compounds e.g., alkaloids, terpenoids, phenyl propanoids, etc.

Pharmaceuticals

Plant secondary metabolites can be divided into several groups which are given below:

Alkaloids

A variety of alkaloids, most of which are plant metabolites, have been used as pharmaceuticals and most of them are plant metabolites.The typical tropane alkaloids, atropine, hyoscyamine, scopolamine and cocaine, were widely used as blockers of the parasympathetic nervous system, such as anodyne and antispasmodic. Various approaches to increase the productivity of alkaloids have been tried by many researchers Vinblastine, an antitumor alkaloid, is most likely to be produced commercially by a Japanese company using a combination process of plant cell culture and that of chemical synthesis.

Morphinan Alkaloids

Papaver somniferum L. (opium poppy) is a traditional commercial source of codeine, and morphine which can be converted to codeine an analgesic and cough-suppressing drug. Mature capsules of *P. bracteatum* accumulates up to 3.5 per cent of thebaine which also can be converted to codeine.

Researchers point out that morphogenetic differentiation from cultured cells of *P. bracteatum* is a prerequisite for producing higher levels of thebaine. Cells of *P. somniferum* produce norsanguinarine, sanguinarine, cryptopine and other various alkaloids, but not codeine and morphine.

Fungal mycelium of *Botrytis sp*. elicits production of sanguinarine by *P. somniferum* cells, increasing the level of this alkaloid 26 times in the presence of the elicitor, 29 per cent of the dry cell weight, compared to the medium without it. The alkaloid extracted from intact plants is added to toothpastes as an anti-plaque agent, but the production of sanguinarine by cell culture technology has not yet been commercialised.

Since no success was gained by the efficient production of thebaine and codeine using cell culture systems by de novo synthesis experimentation was done with the biotransformation of codeinone to codeine using immobilised cells of *P. somniferum*. The conversion yield was 70.4 per cent and about 88 per cent of codeine converted was excreted into the medium.

Berberine

Berberine, us used for intestinal disorders in eastern countries, is an isoquinoline alkaloid which is distributed in roots of *Coptis japonica* and cortex of *Phellondendron amurense*. It takes 5 to 6 years to produce Coptis roots as the raw material.

Researchers find that addition of 10-8 M gibberellic acid into the medium stimulates berberine productivity up to 1.66 g per L of the medium. Using a cell sorter and protoplasts of *C. japonica*, many higher alkaloid-producing cell lines are selected. A "high-density cell culture" process is used to produce berberine much more efficiently now. In order to achieve a cell mass of 70 g/L on a dry weight basis, stirring without damaging the cells, supply of sufficient amounts of oxygen, and that of appropriate nutrients are optimised. As a result, the yields have reached 70 g per L of cell mass and 0.45 g/day of berberine. The continuous culture with high cell density was also been conducted successfully.

Addition of a polyamine, spermidine stimulates the production of berberine by Thalictrum minus cell suspension cultures although other polyamines such as cadaverine, putrescine and spermine are ineffective. Spermidine effects an increase of ethylene generation which is associated with activation of berberine synthesis. The maximum stimulative effect is obtained by the addition of 2 mM spermidine.

Tropane Alkaloids

Scopolamine and hyoscyamine, occurring in the leaves of (*Solanaceae*) plants including *D. myoporoides* and *D. leichhardtii* are being used commercially as anesthetic and antispasmodic drugs. Scopolia, Atropa, Hyoscyamus and Datura also contain tropane alkaloids.

The concentrations of scopolamine and hyoscyamine in cultured cells are generally very low in spite of many efforts to increase the yield the plant cell culture has not yet been employed to manufacture these alkaloids.

Researchers report that roots differentiated from cultured cells accumulate scopolamine, hyoscyamine and/or nicotine, but that the alkaloids are not accumulated in leaves of the regenerated plantlets.

Since the alkaloids are synthesized in the root, cultivated hairy roots transformed with Agrobacterium rhizogenes are cultivated, producing hyoscyamine and other alkaloids at the similar levels to the normal roots.

Agricultural Drugs

Plant Virus Inhibitors

On examining the inhibitory effects of a large number of chemically synthesized compounds and natural molecules on plant viruses it is found that some of them have potent activity as protectors against virus infection.

In order to screen high producing cultured cells of plant viral inhibitors, a variety of callus extracts were examined for the inhibitory activity towards tobacco mosaic virus (TMV) infection. Among these extracts *Phytolacca americana* callus was selected as the most potent producer of the inhibitor. The level of the inhibitor accumulated in the suspension cultured cells which *P. americana* was homogenised and the supernatant was diluted up to 100 times with water. The diluted solution was found to inhibit TMV infection on tobacco and tomato plants significantly.

Using Agrostemma the growth of suspension cultured cells of *A. githago* was somewhat faster than *P. americana.* With Column-lite chromatography and electrofocusing Active principles of *P. americana* and *A. githago* were isolated. At least four basic proteins were obtained from *P. americana* cells whose molecular weights were $1.10X10^4$ to $3.15X10^4$. Among them the highest molecular weight component contained sugars in the molecule. On the other hand, only one basic protein was isolated as the principle from *A. githago* culture and its molecular weight was 2.5×10^4. The proteins obtained from *P.americana* have been widely investigated because of their activities to HIV.

After screening various plants *Mirabilis jalapa* was selected as a producer of an anti-plant virus protein, procuring the callus from leaves of *M. jalapa*. Its suspension cultured cells established were found to accumulate the protein intracellularly. Optimisation of the production and the cell growth as well as selection of high producing cell lines were conducted extensively. The molecular weight of one of the lines was 24 KD and the amino acid sequence was determined which had 24 per cent homology with a ribosome-inactivating protein, ricin D-A chain.

Food Additives

Pigments

In traditional dyeing shikonin and its derivatives such as acetyl shikonin and isobutyl shikonin accumulated in the roots of *Lithospermum erythrorhizon.* The reddish purple pigments used and as a herbal medicine. The plant has also been used as a herbal medicine. Mass cultivation of *L. erythrorhizon* cells produced shikonin compounds to offset the shortage of this plant. Its culture conditions were optimised extensively to increase the level of the products using flasks and various types of fermentors including a rotating cylindrical fermentor.

While *L. erythrorhizon* produced shikonins in White's medium the cell growth was poor in the same medium, whereas Linsmair-Skoog's (LS) medium was recognised to support the growth but not shikonin production.

Therefore, a two-stage culture for mass production of shikonin compounds was used. Namely, to proliferate the cells, LS medium was used at the first stage of the fermentation, and then the cells were transferred into White's medium for production of shikonin compounds. In order to improve the yields further, optimisation of components in both media was carried out extensively, and MG-5 and MG-9 media were established.

Most cultured cells in liquid and solid media occur as aggregates, and selection of high-producing cell lines from the aggregated cells is not effective and labour-intensive. Protoplasts from the cultured cells with appropriate enzymes and selected high shikonin compounds produced protoplasts using a cell sorter. The selected protoplasts were generated to cell lines and cultivated in suspension. From 48 cell lines, a cell line having 1.8 fold the productivity of the parent line was obtained. The cell line showed stable production of shikonin compounds.

To produce the compounds more efficiently, a high-cell density culture process in the second stage of the two-stage cultures. By feeding the nutrients into M-9 medium, the level of cell mass increased and that of the compounds produced increase twofold as much as without feeding. Shikonin and its derivatives are used for lipsticks.

The hairy root culture of *L. erythrorhizon* with Agrobacterium rhizogenes did not produce shikonin on solid MS medium but produced the pigment in the root culture medium and also secreted it into the medium. Addition of absorbents XAD-2, XAD-4, charcoal and so on increased the concentration of shikonin produced.

Anthocyanins

Anthocyanins, the water-soluble pigments responsible for the bright colours in flowers and fruit, are composed of an aglycone (anthocyanidin) and more than one sugar moiety. They normally change colour over the pH range due to the existence of four pH-dependent forms. Thus, at low pH they are red and at higher pH value (over 6) they turn blue. Soft drinks, sugar confectionary, jams and bakery toppings have anthocyanins as an ingredient to give them a red colour. The major source of anthocyanins for commercial purposes are grape pomaces and waste from juice and wine industries, but other potential sources have been investigated. Crude preparations of anthocyanins instead of pure ones are used extensively in the food industry as they are rather inexpensive.

Most of them seem to use an anthocyanin-producing cell line as a model system for secondary product production because of their colour which allows production to be easily visualised.

Accumulation of anthocyanins is enhanced by a high osmotic potential in *Vitis vinifera L.* (grape) cell suspension cultures. Sucrose or mannitol is added in the medium to increase the osmotic pressure and the level of anthocyanins accumulated.

Safflower Yellow

This yellow pigment obtained from the floret of the safflower plant (*Carthamus tinctorius L.*), also known as Mexican saffron or American saffron, has no relation to genuine saffron. The major pigment is carthamin, which exists in the flowers, and there is also a red pigment. Carthamin is the quinoid form of isocarthamin. Safflower yellow is stable to heat and light and is used in baked goods and beverages.

Carthamin is produced from safflower callus cultures. The callus, obtained from flower bud explants, and can also be put into suspension. Medium optimisation has been performed. The production of alpha-tocopherol has been described for safflower cultures. Selection with various media components and precursor feeding experiments have enhanced the production. Addition of polysaccharides, like cellulose, chitin or chitosan increased production of the red pigment. Addition of 1 mM Dphenylalanine and removal of Mg alone or both Mg and Ca from the culture medium also increased the production.

Saffron

Saffron made from the stamens of *Crocus sativus* are prized for their use as a flavouring and colourant and used in baked goods, soups, meat and curry products, cheese, confectionary and as a condiment for the rice of Spanish and Indian foods. The stigma of the plant contains crocin (yellow pigment), safranal (a fragrance) and picocrocin (bitter substance).

The water-soluble crocin, being a glycoside, is not soluble in oils and fats. Saffron is sensitive to pH changes and is unstable towards light and oxidative conditions, but it is moderately resistant to heat. It is used It is also reputed to have medicinal value for stomach ailments.

Since the life of the saffron flowers is very short, making harvesting difficult, it is an ideal target for plant tissue cultures. This problem was tackled through the propagation of saffron stigma-like structures *in vitro*. Further studies show that crocin and picrocrocin are present and, after heat treatment (as done with field-grown stigmas), safranal is produced. The composition of these phytochemicals correspond with that of similarly-treated young, intact stigmas.

References

Bhojwani, S. S., Razdan M. K. *Plant tissue culture: theory and practice, Revised edition* Elsevier, 1996.

Conger, B.V. *Cloning agricultural plants via in vitro techniques.* CRC Press Inc. 1981.

Debergh Zimmerman, (ed.) *Micropropagation: technology and application.* Kluwer Academic Publishers. 1991.

George, E. F. Hall, M. A. and G.-J. De Klerk (eds). *Plant propagation by tissue culture. Volume 1. The background. 3rd edition* Springer, Dordrecht, 2008.

Murashige, T. "Plant propagation through tissue culture". *Annual Reviews Plant Physiology* 25: 135-166. 1974.

Wilkins, C.P., Dodds, J.H. "The application of tissue culture techniques to plant genetic conservation". *Science Progress* 68: 281-307. 1982.

3

Plant Genetic Engineering

The improvement of crops with the use of genetics has been occurring for years. Traditionally, crop improvement was accomplished by selecting the best looking plants/seeds and saving them to plant for the next year's crop.

Once the science of genetics became better understood, plant breeders used what they knew about the genes of a plant to select for specific desirable traits. This type of genetic modification, called traditional plant breeding, modifies the genetic composition of plants by making crosses and selecting new superior genotype combinations. Traditional plant breeding has been going on for hundreds of years and is still commonly used today.

Plant breeding is an important tool, but has limitations. First, breeding can only be done between two plants that can sexually mate with each other. This limits the new traits that can be added to those that already exist in that species. Second, when plants are mated, (crossed), many traits are transferred along with the trait of interest including traits with undesirable effects on yield potential.

Genetic engineering is a new type of genetic modification. It is the purposeful addition of a foreign gene or genes to the genome of an organism. A gene holds information that will give the organism a trait. Genetic engineering is not bound by the limitations of traditional plant breeding. Genetic engineering physically removes the DNA from one organism and transfers the gene(s) for one or a few traits into another. Since crossing is not necessary, the 'sexual' barrier between species is overcome. Therefore, traits from any living organism can be transferred into a plant. This method is also more specific in that a single trait can be added to a plant.

What is Genetic engineering ?

Genetic engineering, also called genetic modification, is the direct manipulation of an organism's genome using biotechnology. New DNA may be inserted in the host genome by first isolating and copying the genetic material of interest using molecular cloning methods to generate a DNA sequence, or by synthesizing the DNA, and then inserting this construct into the host organism. Genes may be removed, or "knocked out", using a nuclease. Gene targeting is a different technique that uses homologous recombination to change an endogenous gene, and can be used to delete a gene, remove exons, add a gene, or introduce point mutations.

An organism that is generated through genetic engineering is considered to be a genetically modified organism (GMO). The first GMOs were bacteria in 1973; GM mice were generated in 1974. Insulin-producing bacteria were commercialized in 1982 and genetically modified food has been sold since 1994.

Genetic engineering techniques have been applied in numerous fields including research, agriculture, industrial biotechnology, and medicine. Enzymes used in laundry detergent and medicines such as insulin and human growth hormone are now manufactured in GM cells, experimental GM cell lines and GM animals such as mice or zebrafish are being used for research purposes, and genetically modified crops have been commercialized.

This article focuses on history and methods of genetic engineering, and on applications of genetic engineering and of genetically modified organisms (GMOs). The article on GMOs focuses on what organisms have been genetically engineered and for what purposes. The two articles cover much of the same ground but with different organizations (sorted by application in this article; sorted by organism in the other). There are separate articles on genetically modified crops, genetically modified food, regulation of the release of genetic modified organisms, and controversies.

Genetic engineering alters the genetic makeup of an organism using techniques that remove heritable material or that introduce DNA prepared outside the organism either directly into the host or into a cell that is then fused or hybridized with the host. This involves using recombinant nucleic acid (DNA or RNA) techniques to form new combinations of heritable genetic material followed by the incorporation of that material either indirectly through a vector system or directly through micro-injection, macro-injection and micro-encapsulation techniques.

Genetic engineering does not include traditional animal and plant breeding, in vitro fertilisation, induction of polyploidy, mutagenesis and cell fusion techniques that do not use recombinant nucleic acids or a genetically modified organism in the process. Cloning and stem cell research, although not considered genetic engineering, are closely related and genetic engineering can be used within them. Synthetic biology is an emerging discipline that takes genetic engineering a step further by introducing artificially synthesized genetic material from raw materials into an organism.

If genetic material from another species is added to the host, the resulting organism is called transgenic. If genetic material from the same species or a species that can naturally breed with the host is used the resulting organism is called cisgenic.

Genetic engineering can also be used to remove genetic material from the target organism, creating a gene knockout organism. In Europe genetic modification is synonymous with genetic engineering while within the United States of America it can also refer to conventional breeding methods. Within the scientific community, the term genetic engineering is not commonly used; more specific terms such as transgenic are preferred.

Plants, animals or micro organisms that have changed through genetic engineering are termed genetically modified organisms or GMOs. Bacteria were the first organisms to be genetically modified. Plasmid DNA containing new genes can be inserted into the bacterial cell and the bacteria will then express those genes. These genes can code for medicines or enzymes that process food and other substrates. Plants have been modified for insect protection, herbicide resistance, virus resistance, enhanced nutrition, tolerance to environmental pressures and the production of edible vaccines. Most commercialised GMO's are insect resistant and/or herbicide tolerant crop plants.

Genetically modified animals have been used for research, model animals and the production of agricultural or pharmaceutical products. They include animals with genes knocked out, increased susceptibility to disease, hormones for extra growth and the ability to express proteins in their milk.

History of Genetic Engineering

Humans have altered the genomes of species for thousands of years through artificial selection and more recently mutagenesis. Genetic engineering as the direct manipulation of DNA by humans outside breeding and mutations has only existed since the 1970s. The term "genetic engineering" was first coined by Jack Williamson in his science fiction novel Dragon's Island, published in 1951, one year before DNA's role in heredity was confirmed by Alfred Hershey and Martha Chase, and two years before James Watson and Francis Crick showed that the DNA molecule has a double-helix structure.

In 1972 Paul Berg created the first recombinant DNA molecules by combining DNA from the monkey virus SV40 with that of the lambda virus. In 1973 Herbert Boyer and Stanley Cohen created the first transgenic organism by inserting antibiotic resistance genes into the plasmid of an E. coli bacterium. A year later Rudolf Jaenisch created a transgenic mouse by introducing foreign DNA into its embryo, making it the world's first transgenic animal. These achievements led to concerns in the scientific community about potential risks from genetic engineering, which were first discussed in depth at the Asilomar Conference in 1975. One of the main recommendations from this meeting was that government oversight of recombinant DNA research should be established until the technology was deemed safe.

In 1976 Genentech, the first genetic engineering company was founded by Herbert Boyer and Robert Swanson and a year later and the company produced a human protein (somatostatin) in E.coli. Genentech announced the production of genetically engineered human insulin in 1978. In 1980, the U.S. Supreme Court in the Diamond v. Chakrabarty case ruled that genetically altered life could be patented. The insulin produced by bacteria, branded humulin, was approved for release by the Food and Drug Administration in 1982.

In the 1970s graduate student Stephen Lindow of the University of Wisconsin–Madison with D.C. Arny and C. Upper found a bacterium he identified as P. syringae that played a role in ice nucleation and in 1977, he discovered a mutant ice-minus strain. He was later successful at created a recombinant ice-minus strain. In 1983, a biotech company, Advanced Genetic Sciences (AGS) applied for U.S. government authorization to perform field tests with the ice-minus strain of P. syringae to protect crops from frost, but environmental groups and protestors delayed the field tests for four years with legal challenges. In 1987, the ice-minus strain of P. syringae became the first genetically modified organism (GMO) to be released into the environment when a strawberry field and a potato field in California were sprayed with it. Both test fields were attacked by activist groups the night before the tests occurred: "The world's first trial site attracted the world's first field trasher".

The first field trials of genetically engineered plants occurred in France and the USA in 1986, tobacco plants were engineered to be resistant to herbicides. The People's Republic of China was the first country to commercialize transgenic plants, introducing a virus-resistant tobacco in 1992. In 1994 Calgene attained approval to commercially release the Flavr Savr tomato, a tomato engineered to have a longer shelf life. In 1994, the European Union approved tobacco engineered to be resistant to the herbicide bromoxynil, making it the first genetically engineered crop commercialized in Europe. In 1995, Bt Potato was approved safe by the Environmental Protection Agency, after having been approved by the FDA, making it the first pesticide producing crop to be approved in the USA. In 2009 11 transgenic crops were grown commercially in 25 countries, the largest of which by area grown were the USA, Brazil, Argentina, India, Canada, China, Paraguay and South Africa.

In the late 1980s and early 1990s, guidance on assessing the safety of genetically engineered plants and food emerged from organizations including the FAO and WHO.

In 2010, scientists at the J. Craig Venter Institute, announced that they had created the first synthetic bacterial genome, and added it to a cell containing no DNA. The resulting bacterium, named Synthia, was the world's first synthetic life form.

Techniques of Genetic Engineering

The first step is to choose and isolate the gene that will be inserted into the genetically modified organism. Presently, most genes transferred into plants provide protection against insects or tolerance to herbicides. In animals the majority of genes used are growth hormone genes. The gene can be isolated using restriction enzymes to cut DNA into fragments and gel electrophoresis to separate them out according to length. Polymerase chain reaction (PCR) can also be used to amplify up a gene segment, which can then be isolated through gel electrophoresis. If the chosen gene or the donor organism's genome has been well studied it may be present in a genetic library. If the DNA sequence is known, but no copies of the gene are available, it can be artificially synthesized.

The gene to be inserted into the genetically modified organism must be combined with other genetic elements in order for it to work properly. The gene can also be modified at this stage for better expression or effectiveness. As well as the gene to be inserted most constructs contain a promoter and terminator region as well as a selectable marker gene. The promoter region initiates transcription of the gene and can be used to control the location and level of gene expression, while the terminator region ends transcription. The selectable marker, which in most cases confers antibiotic resistance to the organism it is expressed in, is needed to determine which cells are transformed with the new gene. The constructs are made using recombinant DNA techniques, such as restriction digests, ligations and molecular cloning. The manipulation of the DNA generally occurs within a plasmid.

The most common form of genetic engineering involves inserting new genetic material randomly within the host genome. Other techniques allow new genetic material to be inserted at a specific location in the host genome or generate mutations at desired genomic loci capable of knocking out endogenous genes. The technique of gene targeting uses homologous recombination to target desired changes to a specific endogenous gene. This tends to occur at a relatively low frequency in plants and animals and generally requires the use of selectable markers. The frequency of gene targeting can be greatly enhanced with the use of engineered nucleases such as zinc finger nucleases, engineered homing endonucleases, or nucleases created from TAL effectors. In addition to enhancing gene targeting, engineered nucleases can also be used to introduce mutations at endogenous genes that generate a gene knockout.

About 1% of bacteria are naturally able to take up foreign DNA but it can also be induced in other bacteria. Stressing the bacteria for example, with a heat shock or an electric shock, can make the cell membrane permeable to DNA that may then incorporate into their genome or exist as extrachromosomal DNA. DNA is generally inserted into animal cells using microinjection, where it can be injected through the cells nuclear envelope directly into the nucleus or through the use of viral vectors. In plants the DNA is generally inserted using Agrobacterium-mediated recombination or biolistics.

In Agrobacterium-mediated recombination the plasmid construct contains T-DNA, DNA which is responsible for insertion of the DNA into the host plants genome. This plasmid is transformed into Agrobacterium that contains no plasmids and then plant cells are infected. The Agrobacterium will then naturally insert the genetic material into the plant cells. In biolistics transformation particles of gold or tungsten are coated with DNA and then shot into young plant cells or plant embryos. Some genetic material will enter the cells and transform them. This method can be used on plants that are not susceptible to Agrobacterium infection and also allows transformation of plant plastids. Another transformation method for plant and animal cells is electroporation. Electroporation involves subjecting the plant or animal cell to an electric shock, which can make the cell membrane permeable to plasmid DNA. In some cases the electroporated cells will incorporate the DNA into their genome. Due to the damage caused to the cells and DNA the transformation efficiency of biolistics and electroporation is lower than agrobacterial mediated transformation and microinjection.

As often only a single cell is transformed with genetic material the organism must be regenerated from that single cell. As bacteria consist of a single cell and reproduce clonally regeneration is not necessary. In plants this is accomplished through the use of tissue culture. Each plant species has different requirements for successful regeneration through tissue culture. If successful an adult plant is produced that contains the transgene in every cell. In animals it is necessary to ensure that the inserted DNA is present in the embryonic stem cells. Selectable markers are used to easily differentiate transformed from untransformed cells. These markers are usually present in the transgenic organism, although a number of strategies have been developed that can remove the selectable marker from the mature transgenic plant. When the offspring is produced they can be screened for the presence of the gene. All offspring from the first generation will be heterozygous for the inserted gene and must be mated together to produce a homozygous animal.

Further testing uses PCR, Southern hybridization, and DNA sequencing is conducted to confirm that an organism contains the new gene. These tests can also confirm the chromosomal location and copy number of the inserted gene. The presence of the gene does not guarantee it will be expressed at appropriate levels in the target tissue so methods that look for and measure the gene products (RNA and protein) are also used. These include northern hybridization, quantitative RT-PCR, Western blot, immunofluorescence, ELISA and phenotypic analysis. For stable transformation the gene should be passed to the offspring in a Mendelian inheritance pattern, so the organism's offspring are also studied.

Genetic Transformation Techniques

In both basic and applied molecular biology DNA can be moved into an organism, thereby altering its genotype or genetic make-up. Genes derived from unrelated species and even other kingdoms, such as bacteria, fungi, plants, animals, that would otherwise be inaccessible to an organism, can be combined in the lab using genetic transformation techniques.

In a laboratory transformation involves *Escherichia coli* or *Saccharomyces cerivisiae* and a thermal or electrical stimulus. In the pharmaceutical industry, bacteria and yeast are transformed with selected human genes to produce therapeutic benefits to treat human disorders. Most of the insulin now supplied to diabetics is produced this way.

Conjugation takes place in nature, whereby many strains of bacteria exchange genetic information and the new genetic information is passed on to subsequent generations. The single-celled nature of the organisms enables the change of only one cell in order to integrate the new genetic information and allow for its transmission to the next generation. This natural exchange of genetic information allows the organism an increased capacity to adapt to its environment.

In multicellular organisms transformation is usually more difficult, particularly with some important plant species like cereals. The multicellular nature of most plants poses the problem of transforming each cell of the plant in order to fully integrate the new information. The transformation of an individual plant cell and then regenerating it into a whole organism calls upon the pluripotent nature of plant cells.

Bacteria help to genetically transform plants and other organisms. Viruses are also able to move DNA (or RNA) into an organism and cause immense changes. Soil bacteria, such as *Agrobacterium tumefaciens* and *Agrobacterium rhizogenes*, are great examples of natural transformation systems, causing crown gall disease and 'hairy root syndrome' respectively. *Agrobacterium* can transfer a section of its own DNA, known as the T-DNA or 'transfer DNA', into plant cells. For *A. tumefaciens* this normally occurs at wound sites and the transferred DNA causes the plant to develop crown gall disease. Apart from causing a gall, the *Agrobacterium* harnesses the plant's machinery to produce unique sugars called opines that the bacterium uses as a nutrient supply.

Scientists have used one of the most common methods for introduction of DNA into plant cells using 'disarmed' *Agrobacterium tumefaciens* – which the DNA of interest has replaced the disease genes. DNA of interest can be any valuable agricultural trait, such as genes that provide herbicide tolerance in canola and other crops. The particle bombardment or biolistics transformation method uses rapidly propelled tungsten microprojectiles coated with DNA that are 'fired' into plant cells. The DNA of interest is often incorporated at random into the plant DNA. Other methods, such as electroporation and microinjection, have also been employed to transform plant cells.

Cells that have been successfully transformed are selected and the DNA of interest is typically cloned adjacent to DNA for a selectable marker gene such as *nptII* (encoding kanamycin antibiotic resistance). The transformed plant cells are then cultured in the presence of the antibiotic and only cells or tissues expressing the new DNA survive. In order to eliminate or exclude the antibiotic resistance markers in transformation experiments, effective methods have been developed. Fluorescence genes such as GFP (green fluorescent protein) can also be used to report transformation events.

The transfer and expression of foreign genes into plant cells has been one of the most recent and successful areas of plant cell and tissue culture developed for exploring potential benefits in agriculture. The transformation of bacterial cells by uptake of DNA from another bacterium, and subsequent incorporation of this foreign DNA into the genetic material of the host is well established. However, genetic modification of higher plants by introducing DNA into their cells is a highly complex process, involving a foreign nucleotide sequence that has to be introduced in such a way that allows expression of alien genetic information in the transformed cells.

Availability of restriction endonucleases, development of the DNA-DNA hybridisation technique and marker genes for selection of transformed cells have now enabled integration of exogenous DNA into cells of Monera, fungi, animals, and higher plants. The addition of new genetic information as a result of these processes in eukaryotic plants expresses at the cellular and later, at the whole organism level, and can also be transmitted in successive generations of transformed individuals. Such transgenic plants, since normally they originate following an exchange of genetic information between sexually incompatible species, are not known to exist in nature.

Using whole plant systems such as seeds and seedlings genetic modification of plants was done by introducing isolated DNA. Exogenously applied DNA is taken up by plants, and integrated into the host genome, based mainly on isopycnic centrifugation in cesium chloride (CsCl) gradients.

Integration and replication are two major types of experiments that were conducted. In the former radioactive labelled DNA of density different from the host DNA was administered to the seedlings, followed by isolating DNA from seedlings, or their excised parts, after a period of metabolism for analysis on CsCl gradients. The radioactive donor DNA was ascertained for integration by the occurrence of a radioactive band approximately intermediate in density between the host and donor DNA. The intermediate density band was further tested by sonication and denaturation. The replication experiments were also conducted in a similar manner and differed only in respect of the donor DNA, which was non-radioactive. These experiments also showed proof of bacterial plant DNA complexes in tomato, barley and Arabidopsis.

Trying transformation experiments using a similar donor DNA with the same host or different plant systems have not yielded the same results, showing that there was neither integration nor replication of the exogenous DNA inside the host cells. In fact, the donor DNA ultimately degraded in the host and the degraded products were used for endogenous DNA synthesis.

Genetic transformation of whole plant systems seems to have achieved success with cereals. There has been much interest in the reported rhizobial infection of non-legumes to form an effective nodular nitrogen-fixing symbiosis within the dicot plants belonging to the genus Parasponia so much as to extend nodular symbiotic association to monocots, especially the cereals. Cocking and co-workers reported induction of nodular structures on the roots of rice seedlings when these roots were treated with a cellulasepectolyase enzyme mixture followed by inoculation with either

Rhizobium, or Bradyrhizobium, in the presence of PEG. This has opened new vistas in the genetic engineering of plants for possibly extending this technique for effective nodulation in respect of other cereals as well as non-legumes. It has to be seen whether nodular structures on transformed rice are effective in nitrogen fixation. Otherwise, other strains of genetically engineered Rhizobium, or its wild type, may be needed for this purpose.

Uptake of DNA by Pollen

In 1976, researchers incubated *Nicotiana glauca* pollen with DNA isolated from *Nicotiana langsdorffii* and then used this pollen to pollinate emasculated *N. glauca* flowers. This experiment led to the production of transformed plants with developed tumours at the wound site on the stem, characteristic feature exhibited by the sexual hybrids arising between these two species. Pollen from maize plant homozygous for aleurone and various endosperm marker colours were mixed with DNA extracted from a strain homozygous dominant for all these traits yielding a high pollen transformation efficiency. These observations suggest that pollen may act as vehicles for the transfer of foreign genes into plants.

Young floral tillers of rye plant were manually injected with a solution of plasmid DNA carrying a chimeric gene having T-DNA nopaline synthase gene flanking coding sequence of the Tn 5 neo gene (expressing for antibiotic aminoglycoside resistance). These Injected flowers were pollinated with normal pollen and progeny seeds plated on kanamycin medium. Of the 3,000 seeds, only two seedlings were isolated showing traits of kanamycin resistance and the nos-neo gene.

Transformation of Protoplasts

Genetic modification of cells has benefited by the isolated protoplast system which proved most rewarding. Since there is no wall around the plasma membrane, protoplasts are able to fuse and can also take up chloroplasts, nuclei, micro-organisms and isolated foreign DNA. This has induced researchers to investigate various modes of genetic modifications in cells using the protoplast system.

Uptake of Organelles

Chloroplast Transplantation

The uptake of chloroplasts by albino protoplasts isolated from petunia, tobacco and carrot plants conducted in experiments required a simple treatment exposing a mixture of albino protoplasts and chloroplasts to PEG in a manner similar to protoplast fusion. PEG caused fusion of the two external membranes of the plastids and while entering the protoplasts both the plastid and protoplast membranes fused, causing disruption of these organelles. Hence, focus was diverted from direct uptake to transfer of chloroplasts through the fusion of protoplasts, making it possible to restore photo-autotrophy in albino protoplasts by fusion with wild type (green) protoplasts in a number of species. A gamma-irradiated chloroplast-mutant of Nicotiana species was used for interspecific chloroplast transfer by protoplast fusion, producing cybrids, whereas protoplast fusion was suitable to rescue the chloroplast mutants of *N. plumbaginifolia* from undesirable nuclear backgrounds.

Cloroplast transformation via protoplast fusion in plant improvement can occur by transfer of genes for herbicide resistance between related species. For example, isolation of mutants by cell selection in populations of protoplast-derived colonies yield triazine-resistant Nicotiana plants. The resistant trait arises as a result of the mutation in the chloroplast gene psb-A. Similarly, researchers found that inhibitors of bacterial protein synthesis were used to induce chloroplast mutations in protoplast cultures of *N. plumbaginifolia* making it possible to isolate mutants resistant to antibiotics and photosynthesis-inhibiting herbicides. The bacterial APH (3) II gene from transposon Tn 5 and the cat gene from transposon Tn 9 are selectable marker or reporter genes that have been found suitable for transfer of mutant chloroplast genes.

Control of light-mediated proteins expressed by rbcs and cab gene families can be achieved by the interaction between chloroplast genes from weeds to sexually incompatible crop plants. This is another area that may have potential in future since it has played a useful role in understanding the nuclear-chloroplast interaction, physiology of isolated plastids and transfer of herbicide resistance.

Mitochondrial Transplantation

Though there is no evidence regarding the direct uptake of isolated mitochondria by protoplasts there are however reports on incorporation of mitochondrial DNA through protoplast fusion leading to the formation of cybrids. It may also be possible to transfer traits such as toxin sensitivity of Texas male sterile cytoplasm encoded by mt DNA in maize.

Nuclear Transplantation

The methods for isolation and uptake of nuclei by mesophyll protoplasts of petunia, tobacco and maize have been refined to facilitate expression of the transferred genome in the cells. Vicia hajastana nuclei in protoplasts of an auxotrophic cell line of Datura innoxia were transplanted and prototrophic cell colonies expressing for Vicia genomic DNA in cultures were isolated. However, these colonies on an appropriate regeneration medium developed shoots typical of wild type Datura.

The uptake of isolated chromosomes, instead of complete nuclei, Holds the potential for transfer of foreign genes into plant cells. Development of the flow cytometry technique permitted the introduction of isolated chromosomes into protoplasts of Haplopappus, Nicotiana and Petunia. Only metaphase chromosomes were found suitable. Isolated chromosomes can also be introduced into protoplasts by microinjunction. In most of these experiments transplanted chromosomes were eliminated due to incompatibility.

Experiments to transfer micronuclei into protoplasts were also conducted. Amiprophos-methyl (APM) treatment of potato and carrot cells encouraged accumulation of a large number of metaphases. The dispersed single or groups of chromosomes in these cells acquired a nuclear membrane and formed micronu-clei which could be separated by flow cytometry and then used for transplantation. This widens the application prospect of micronuclei for transferring specific intact chromosomes for the purpose of gene mapping.

Uptake of Microorganisms

There were trial to transfer whole bacteria (Rhizobium) into protoplasts for utilising the transformed protoplasts for nitrogen fixation but there yielded little success. Therefore, the use of viruses as vectors for genetic engineering studies gained focus. The mechanism of infection, nuclear-protein interaction, integration and expression of viral DNA in isolated protoplasts should be possible since in nature viruses do infect plants and multiply within their tissues. Earlier transducing phages were used to transfer bacterial genes into plant cell lines or callus tissues but there was no evidence of "transgenosis" in long term cultures.

Experiments on uptake of TMV by isolated protoplasts since 1969 have shown the incorporation of viral nucleic acids into protoplasts, using them as vectors in genetic transformation. Most plant viruses have encapsulated genomes, assuming that no specific virus receptor sites exist on the exterior of the plasma membrane. Viral based transforming agents may be of the integrating or non-integrating type.

The following factors are generally considered in selecting the progenitor for the transforming agent: (a) amenability to genetic engineering techniques, (b) packaging constraints in the particle and (c) development of a system in which expression can be studied. The most amenable and widely chosen progenitor for genetic engineering studies is the cauliflower mosaic virus (CaMV). Moreover, the dsDNA genome in CaMV has appropriate restriction sites compatible with the methodology developed for *Agrobacterium tumefaciens* plasmid. A chimeral CaMV carrying methotrexate resistance was successfully introduced into turnip cells in such a way that the virus replicated normally in the transformed plants. Similarly, barley protoplasts infected with the ssRNA bromovirus (BMV), on having the chloramphenicol acetyltransferase (CAT) bacterial gene inserted into the RNA3 genome, transcribed for CAT activity. Gemini viruses may also be used for genetic transformations but their usefulness as vectors are still under study.

Agroinfection which involves a combination of viral and bacterial vectors is another approach being considered. Various strategies developed for uptake of viruses by plant protoplasts and their use as transforming agents. Use of fungal protoplasts as eukaryotic gene vectors for plant cell transformation has also been attempted. Fungal protoplasts have to be experimented upon since fungi and higher plants establish a range of nutritional relationships in nature. *Catharanthus roseus* protoplasts were fused with fungal protoplasts as an attempt to transfer the secondary metabolite synthesising ability of fungus into a higher plant protoplast system in cultures. Later, other researchers successfully mediated uptake of celery protoplasts by *Aspergillus nidulans* and *Fusarium oxysporium* protoplasts using a PEG solution, which was confirmed by ultrastructural studies and diamidino-2-phenylindole staining. Protoplasts of celery showed improvement in the level of viability after uptake by *A. nidul ans,* whereas with *F. oxysporium* the level decreased. However, whether uptake of fungal protoplasts can be exploited for secondary metabolite synthesis in higher plant cell or protoplast system is yet to be seen.

Transformation Using Agrobacterium

In the past few years, there have been molecular studies of crown gall disease by Agrobacterium plasmids and the use of recombinant DNA technology, with further investigations using Agrobacterium for plant genetic manipulations.

Agrobacterium tumefaciens, a soil bacterium, can infect most dicotyledonous plants, usually at a wound site. Where the tissue develops a neoplastic growth known as a crown gall tumour. The tumour tissue excised from the plant is capable of growing on hormone-free medium independent of the bacterium. In cultures, the crown gall tissue produces a set of metabolites termed opines. Agrobacterium fosters the Ti plasmid, which is directly responsible for tumour induction. The transfer of small DNA segments from this plasmid and their integration into the genome of host cells generates tumour formation in the plants. The property of the Ti plasmid DNA segment (T-DNA) integrating with the genome of higher plants has boosted the use of Agrobacterium in genetic transformation techniques.

The infection by Agrobacterium rhizogenes another bacterium used in plant genetic manipulations, causes hairy roots which are of clonal origin. Unlike *A. tumefaciens* tumours, the tumour produced by *A. rhizogenes* is capable regenerating mature fertile plants. The tumour-inducing properties of rhizogenes strains are also carried on a large Ri plasmid DNA (Ri T-DNA). Two segments of this DNA (TL-DNA and TR- DNA) are transferred to plant cells during infection by *A. rhizogenes.*

Agrobacterium Mediated Gene Transfer

With the improvement in plant transformation vectors and methodologies the efficiency of plant transformation has increased and to achieve stable expression of transgenes in plants achieved. The simplicity of the transformation system and precise integration of transgenes make it viable for *Agrobacterium* Ti plasmid-based vectors to continue offering the best system for plant transformation. Binary vectors have been improved by the integration of supervirulent *vir* genes, matrix attachment regions (MAR) and the insertion of introns in marker genes and reporter genes. These improvements and the use of acetosyringone, have led to transformation of monocotyledonous plants using *Agrobacterium*. The green fluorescent protein (GFP) gene has been extensively modified for plant codon preference, ER/plastid targeting and for greater solubility thereby making it an adaptable vital reporter for transgenic plants. There is also development of transgenic plants devoid of antibiotic marker genes.

Marker genes in transcenic plants have been eliminated by cotransformation of multiple T-DNAs, site-specific recombination strategies and deployment of Ac/Ds-based transposition. Positive selection strategies using *ipt*, xylose isomerase and phospho-mannose isomerase are shown to be useful in many crop plants. The development of BIBAC vectors, a demonstrated capability to transfer multiple genes of a pathway and successful T-DNA tagging in rice, are evidence to show that transformation technologies can be deployed for the study of 'functional genomics' in plants.

The particle bombardment system still has its use in organelle transformation and transformation of plants that lack efficient regeneration systems. This has led to the design of vectors that minimise transgene silencing while ensuring desired levels of transgene expression. The mechanism of T-DNA integration in plants has to be studied so that 'knock out' mutagenesis and homology-based gene replacements can be achieved in plants.

Plant transformation is performed using a wide range of tools such as *Agrobacterium* Ti plasmid vectors, microprojectile bombardment, microinjection, chemical (PEG) treatment of protoplasts and electroporation of protoplasts. Among all the methods employed transformation using *Agrobacterium* and microprojectile bombardment is currently the most extensively used method. Recent developments in the phenomenon of 'gene silencing' has come to centrestage after a large number of transgenic plants have been carefully evaluated for transgene expression in successive generations.

Agrobacterium tumefaciens have precisely transferred defined DNA sequences to plant cells been very effectively utilised in the design of a range of Ti plasmid-based vectors. Three genetic elements, *Agrobacterium* chromosomal virulence genes (*chv*), T-DNA delimited by a right border and a left border and Ti plasmid virulence genes (*vir*) constitute the T-DNA transfer machinery. The mechanisms governing the transfer of 'T-complex' via the conjugation channel and the roles of plant and *Agrobacterium* proteins in T-DNA integration are being intensely studied.

There are *Agrobacterium*-based DNA transfer system advantages in plant transformation which are:

i) The simplicity of *Agrobacterium* gene transfer makes it a 'poor man's vector.
ii) A precise transfer and integration of DNA sequences with defined ends.
iii) A linked transfer of genes of interest along with the transformation marker.
iv) The higher frequency of stable transformation with many single copy insertions.
v) Reasonably low incidence of transgene silencing.
vi) The ability to transfer long stretches of T-DNA (> 150 kb).

For long, *Agrobacterium* was unable to transfer DNA to monocotyledonous plants but with effective modifications in Ti plasmid vectors and finer modifications of transformation conditions, a number of monocotyledonous plants including rice, wheat, maize, sorghum and barley have now been transformed. In fact, *Agrobacterium* T-DNA transfer is now considered as 'universal' based on successful transformation of yeast, *Aspergillus*and human cells.

Ti Plasmid-based Vector System

Agrobacterium vir Helper Strains

Plant transformation use a typical binary vector system comprising an octopinetype *vir* helper strain, such as LBA4404, that harbours the disarmed Ach5 Ti plasmid and a binary vector such as pBin19. the nopaline-type MP90 and the L, L-succinamopine-type EHA101 have been added to the available range of *vir* helper strains. The bacterial kanamycin resistance gene in EHA101 was deleted to develop the *vir* helper strain EHA105. EHA101 and EHA105, due to harbouring the 'supervirulent' *vir* genes, exhibit show broader host-range and higher transformation efficiency. Many recalcitrant plants, such as rice, wheat and barley have been transformed using EHA101 and EHA105. A new *vir* helper strain pTiChry5 has been constructed from an *Agrobacterium* strain virulent on soyabean. This strain has good potential for legume transformation.

Binary Vectors

Since 1984, many modifications have been made in binary vectors to expand the range of their utility and to improve their transformation efficiency, one such being the construction of the superbinary vector pTOK233 by cloning the *vir*B, *vir*G and *vir*C genes of pTiBo542 in pGA472. This led to the successful transformation of *japonica* rice, *indica* rice, *javanica* rice, maize, *Sorghum* and *Allium cepa*were using the same strategy. A long-held notion that monocots are not amenable to *Agrobacterium*-mediated transformation has been proved wrong. Recently, pTOK233 was used to transform a pulse, moong bean.

Many finer improvements have been made in binary vectors: pBIN19 has been completely sequenced; its improved version, pBIN20, with many additional single restriction sites in the MCS was reported recently; a new series of pPZP vectors have been developed which are small in size and stable in *Agrobacterium*; the pPZP vector backbone was used to construct the pCAMBIA series of vectors with *npt*II, *hpt* or *bar* as selection markers and *gus* or *gfp* as reporters. Plant expression vectors of pRT100 series permit construction of gene cassettes with CaMV 35S promoter and its polyA signal, which can be excised and placed in the MCS of binary vectors.

The problem of availability of single restriction sites, particularly in the expression cassette with the gene of interest, was dealt, with by the construction of a set of plasmids, pART7 and pART27. The shuttle plasmid pART7 has a MCS placed between CaMV 35S promoter and *ocs* polyA signal. The expression cassette is flanked on either side by *Not*I sites. The coding sequence of a gene of interest is cloned in the MCS of pART7. The expression cassette is excised by using *Not*I and cloned in the *Not*I site of the binary vector, pART27 that is designed so as to have the *npt*II gene close to the left border to enables the selection marker is the last to be transferred into plants. This strategy ensures that all plants selected on kanamycin will have a complete TDNA inclusive of the gene of interest.

In functional genomics *Agrobacterium* Ti plasmid vectors have important operations. Random integration of TDNA in the plant genome causes mutations and simultaneously tags the gene into which insertion occurred. T-DNA tagging has cloned many *Arabidopsis* genes. A similar effort in rice resulted in the generation of 22,090 transgenic plants. Phenotypic analysis of these mutants is expected to lead to the discovery of many new gene functions.

Genome studies require binary vectors that can transfer T-DNAs several hundred kbs. Using binary bacterial artificial chromosome vectors (BIBAC) a T-DNA consisting of 150 kb human DNA was precisely transferred and integrated into tobacco genome. These vectors are expected to play a key role in positional cloning of plant genes and in functional genomics.

Engineering of plants with new pathways may require the transfer of multiple genes. The technical difficulty of placing multiple genes in one T-DNA encountered due to loss of single restriction sites in successive cloning steps was overcome in engineering provitamin A biosynthesis in rice. Daffodil *phytoene synthase* gene and *Erwinia phytoene desaturase* gene were placed in one T-DNA. The second T-DNA carried daffodil *lycopene-b-cyclase* and *E. coli hpt*. Interestingly, 12 out of 60 plants selected on hygromycin carried the genes of the second T-DNA as well. Therefore, 20 per cent of hygromycin-resistant plants comprised all the three genes of β-carotene biosynthetic pathway. Thus, multiple genes of a pathway can be transformed by cotransformation of multiple T-DNAs. Though particle bombardment permits cotransformation, *Agrobacterium* offers the additional advantage of precise integration of complete lengths of multiple T-DNAs.

Matrix attachment regions (MARs), the AT-rich DNA sequences that promote the attachment of DNA to the nuclear matrix or inner nuclear membrane, produce a series of loops. MARs have two functions: (a) Genomic compartmentalisation of coexpressed genes in individual loops and (b) Formation of the boundaries of transcriptionally active DNA loop domains, thereby blocking the inhibitory influence of neighbouring sequences. MARs have potential applications in transgenic plants. A transgene flanked on either side by MAR sequences may form an independently expressing loop domain thereby minimising position effects. Researchers constructed *gus* gene flanked by tobacco MAR sequences and transformed cell suspension cultures by particle bombardment, thereby increasing the expression of integrated *gus* gene by 140-fold. *Gus* and *gfp* genes were constructed with the tobacco Rb7 MAR sequences generating transgenic rice plants by particle bombardment. The presence of MAR sequences increased the expression of reporter genes in the range of 3.3-fold to 650-fold. The reduction in reporter gene expression, generally attributed to flanking sequences at the sites of integration, was also minimised.

Improvements in Marker Genes and Reporter Genes

Leaky expression of plant selection marker genes such as *npt*II and *hpt* in *Agrobacterium*, results in bacterial over growth during selection and in the emergence of false-positive plants. To avoid this, researchers placed the intron 2 of the potato ST-1 gene in the N-terminal part of the *npt*II gene. Lack of mRNA splicing in *Agrobacterium* completely eliminated leakiness in kanamycin selection. In another study, the castor bean catalase 1 intron was placed in the *hpt* gene to avoid leakiness in hygromycin selection. The binary vectors with intron-*hpt* significantly improved rice transformation efficiency by improving *hpt* expression and by limiting *Agrobacterium* overgrowth.

The histochemical analysis of expression of *gus* as a reporter gene help in Early detection of transformation events. A major limitation of this approach is the expression of this reporter gene in *Agrobacterium* despite the use of plant promoters. This problem was also tackled by inserting introns, which are processed in plants but not in *Agrobacterium*. The intron 2 of potato ST-LS1 gene was placed in the coding sequence of *gus* gene. In another study, the castor bean catalase gene intron was inserted in the Nterminal part of the coding sequence of *gus*. In both cases, *gus* activity was limited only to transformed tissues and was not detected in *Agrobacterium*.

Transgenic plants make use of green fluorescent protein gene (*gfp*) from jellyfish finds. GFP, unlike *gus*, is foreseen in living cells. Another reporter gene, luciferase (*luc*), requires an externally added substrate for detection. However, GFP fluorescence occurs with uv/blue light and oxygen without any externally added substrate. The first series of transformation experiments showed the following limitations of the application of wild type *gfp* gene in plants: (a) Codon preference was different between jellyfish and plants. (b) Presence of a cryptic intron in *gfp* was recognised and spliced in plants. (c) Low solubility of GFP in cytoplasm. (d) Interference of cytoplasm-accumulated GFP in the regeneration of transgenic shoots. Synthetic versions of *gfp* (e.g. modified *gfp*, *mgfp*) helped to avoid the cryptic intron and changed the codon preference to that of flowering plant genes. Thereby dramatically increasing the intensity of fluorescence in transgenic *Arabidopsis* and tobacco plants, and full length GFP being translated. Elevated expression of GFP unfortunately foiled the regeneration of transgenic shoots. However, fusion of a signal peptide-encoding sequence to *mgfp*4 and consequent targeting of mgfp4–ER protein to ER resulted in routine regeneration of highly fluorescent shoots.

The reduced solubility of GFP in cytoplasm was considered as an important reason for reduced fluorescence. Therefore the synthetic version, *mgfp*4 was subjected to triple sitedirected mutagenesis (F99S, M153T, V164A) to generate a soluble-modified *gfp* (*sm-gfp*). Expression of *sm-gfp*4 in *Arabidopsis* gave brighter fluorescence compared to *mgfp*4, possibly due to increased solubility of *sm-gfp*. Normally, GFP is excited at 396 nm or 475 nm and emits green fluorescence maximally at 508 nm. Experiments carried out with two site-directed mutagenesis experiments on *sm-gfp* generated spectral variants: a soluble-modified red-shifted version of GFP (smRSGFP) and a soluble-modified blue fluorescent version of GFP (sm-BFP).

Using the rbcSSU promoter and transit peptide (rbc-Tp-*sgfp*) plastid-targeted version of synthetic GFP (*sgfp*) was constructed. Transgenic rice plants generated using *Agrobacterium* with rbcS-Tp-*sgfp* showed 20-times higher fluorescence compared to a version without plastid targeting. Two groups placed intron-2 of potato ST-LS1 gene in the synthetic *gfp* versions. In the latter report, the intron was placed in *mgfp*4-ER, sm-*gfp*4, smRS-GFP and sm-BFP genes. Intron splicing occurred efficiently in tobacco and in maize cells.

Modifications in Cocultivation Conditions

Today, it is a common practice to have cocultivation of explants with *Agrobacterium* in the presence of acetosyringone, a *vir* gene inducer, in the transformation of recalcitrant crops such as rice, maize, barley and wheat. Preinduction of *Agrobacterium* with 400 mM acetosyringone prior to cocultivation is important in rice transformation. Finer changes such as supplementation of buffers to cocultivation medium and preincubation of explants that increase *vir* gene induction increase transformation efficiency of *Agrobacterium*. Many physical factors influence transformation efficiency. Uniform wounding of tobacco leaves and sunflower apical meristem by particle bombardment, vacuum infiltration of *Arabidopsis* seedlings, rapid proliferation of rice calli at 30° C and cocultivation at a lower (22° C) temperature led to an increase in the efficiency of *Agrobacterium*-mediated transformation.

'In planta' transformation using *Agrobacterium*

Growth of '*in planta*' transformation by floral dips has come about due to the small size of *Arabidopsis* plant and improved efficiency of transformation by vacuum infiltration. This method obviates plant tissue culture and eliminates somaclonal variations. In a series of subsequent studies, it was found that ovules (female gametophytes) of immature flowers were more the often targets of floral dip transformation of *Arabidopsis.*

Medicago truncatula, an emerging model legume plant, has recently seen the '*In planta*' transformation. Infiltration of flowering plants was superior to infiltration of young seedlings. Transformation by vacuum infiltration used binary vectors with *bar* gene and 'BASTA' spray screened the TI progeny. This is a potentially useful approach in plants like pulses where regeneration protocols are either not available or regenerating parts of plants are not accessible to *Agrobacterium*-mediated transformation.

Marker-free Transgenic Plants

Antibiotic resistance markers (e.g. *npt*II, *hpt*) or herbicide resistance markers (e.g. *bar*) are required for selectively propagating transformed cells and tissues though their subsequent maintenance in transgenic plants is unnecessary. Markers have to be eliminated as the antibiotic resistance genes may be transferred to pathogenic bacteria or the herbicide resistance genes may be transferred to weeds. Besides removal of marker genes offers the following research advantages: It enables (i) successive rounds of transformation so that useful transgenes can be stacked without crossing, (ii) retention of promoters along with selection markers which will lead to the presence of multiple copies of a promoter, thereby activating signals for transcriptional gene silencing.

Marker Elimination by Site-specific Recombination

Site-specific recombination systems have proved beneficial for the elimination of selection markers. In the P1 bacteriophage recombinase (*Cre*) and its recognition site (*lox*P). Researchers cloned the luciferase (*luc*) gene and *lox*P/*hpt*/*lox*P cassette in a binary vector and generated transgenic tobacco plants. Then the transgenic plant with *luc*/*lox*P/*hpt*/ *lox*P was retransformed with a binary vector with *cre*/ *npt*II. In 10 of the 11 kanamycin-resistant plants obtained, excision of *hpt* had occurred. Chances were that *luc* of the first TDNA and *cre*/*npt*II of the second T-DNA were present in unlinked loci. Selfing of the T_0 transgenic plants was done and T_1 plants were analysed. In one case, 17 of 104 T_1 plants had luciferase (*luc*) activity but lacked kanamycin resistance. Thus, in a two-step exercise the marker genes (*hpt* and *npt*II) and the recombinase (*cre*) were eliminated. Instead of retransformation, crossing of *luc*/ *lox*P/*hpt*/*lox*P plants with *cre*/ *npt*II plants also yielded similar results.

In maize cells, the recombination system of the 2 mm plasmid of *Saccharomyces cereviseae* involving FLP recombinase and FRT recombination site has been successfully deployed for marker elimination. A combination of lambda attachment site (*att*P) and negative selection using *tms*2 and napthaleneacetamide (NAM) were used for marker elimination by intrachromosomal recombination.

The maize Ac/Ds system was tested for marker elimination in tomato using a T-DNA with Ds-*gus*-Ds/Ac/*npt*II for transformation. Transposition events in T_0 plants would place *gus* and *npt*II genes in unlinked loci thereby segregating them in T_1 plants. In a multi-auto-transformation vector (MAT) a combination of Ac functions and *ipt* gene were used thereby eliminating *ipt* without selfing and segregation.

A single-step *ipt*-MAT vector for rice incorporated the recombinase (R) and recombination sites (RS) of *Zygosaccharomyces rouxii* whereby the T-DNA constructed by them carried *nptII*/*gus*/*hpt*/RS/*ipt*/35SR/ *gfp*/RS sequences. *Agrobacterium*-mediated transformation of scutellum-derived callus was performed, with selection based on the ability to form normal shoots on cytokinin-minus medium. In 25 per cent of the infected calli, regenerated normal shoots carried the *gus* gene without *ipt*, showing that marker (*ipt*)-free rice plants could be generated in a very short span of one month.

Elimination of Marker Genes by Cotransformation

A simple approach for marker elimination uses cotransformation of two independent T-DNAs, one with selection marker and the other with the gene of interest. If the two T-DNAs integrate at different loci in the T0 plant, they will get segregated in the T_1 progeny leading to the generation of marker-free plants.

Based on the logical idea that conditions which support. Two separate T-DNAs were placed in superbinary vectors that harboured supervirulent *vir*B, *vir*G and *vir*C genes based on, the logical idea that conditions supporting high efficiency of transformation may lead to a high frequency of cotransformation. One T-DNA contained a drug resistance marker and the second T-DNA carried the *gus* gene. There was cotransformation of *gus* in more than 47 per cent of the drug-selected tobacco and rice plants. The non-selected *gus* gene segregated from *npt*II or *hpt* genes in T_1 progenies of more than 25 per cent of drug-selected plants. This simple two-step strategy is a convenient method to generate marker-free transgenic plants.

As an alternate strategy of cotransformation was to place two separate, compatible binary plasmids in one *Agrobacterium*. The selected T-DNA with *npt*II was placed in a RK2 replicon and the non-selected *gus* gene was placed in a pRiHR1 replicon. Cotransformation in T_0 plants and segregation of *npt*II and *gus* in the T_1 plants occurred at a high frequency. Two T-DNAs of variable lengths were placed on the binary plasmid with the *npt*II TDNA 7 kb long and the *hpt* T-DNA 3.2 kb long. The cotransformation frequency of the non-selected T-DNA was higher when the marker on the longer T-DNA (*npt*II) was used for selection. The selected marker could be genetically separated (segregated) in the T_1 plants. This shows that the T-DNA with the selection marker should be twice longer than the non-selected T-DNA to achieve high cotransformation in T_0 plants and to eliminate the marker by segregation in the T_1 plants.

Positive Selection of Transgenic Plants

While research to eliminate undesirable markers from transgenic plants was under way, scientists were making a parallel exploration of the use of positive selection markers. Positive selection markers unlike antibiotic and herbicide markers, which kill the untransformed cells, render metabolic/developmental advantage to transformed cells and help in faster proliferation and regeneration. There is also the elimination of the problem of untransformed dead tissues being detrimental to the growth of transformed tissues. A glucuronide derivative of benzyladenine (BA). β-Glucuronidase, expressed in the transformed tissues, released BA and supported tissue proliferation and regeneration in a cytokinin-minus medium is a fine example of a positive selection agents. The transformation efficiency obtained thus was twice than that obtained with kanamycin. Another strategy for positive selection was based on the inability of plants to use xylose as carbon source unless it is converted to xylulose by xylose isomerase. *Thermoanaerobacterium thermosulfurogenes* xylose isomerase gene (*xyl*A) was successfully used as a positive selection marker in potato and tomato, the transformation efficiency of positive selection using xylose being comparable to that obtained with kanamycin.

With the positive selection method using *E. coli* phosphomannose isomerase gene (*pmi*) now well established, mannose is converted to mannose-6-phosphate by plant hexokinases. Plants lack phosphomannose isomerase. Mannose-6-phosphate that collects in cells, exerts severe growth restraint by inhibiting phosphoglucoisomerase and by depleting phosphate required for ATP synthesis. Mannose is not directly toxic and using is selection transformation of sugarbeet, maize, cassava and rice has been accomplished. Yeast 2-deoxyglucose-6-phosphate phosphatase conferring resistance for 2-deoxyglucose and *Catharanthus roseus* tryptophan decarboxylase that converts the toxic tryptophan analogue 4-methyl-tryptophan to non-toxic 4-methyltryptamine are other tested examples of positive transformation markers.

Problem of 'Long Transfer' of Plasmid Backbone

Agrobacterium T-DNA transfer has the unique advantage of accurately processing the T-DNA between the right and left borders and its precise transfer and integration to the plant genome. However, there are possibilities that non-T-DNA portions may also get transferred to the plant genome. T-DNA transfer initiates in low frequencies from the left borders as well. What is known as 'long-transfer' occurs when, at fairly high frequencies, TDNA transfer starting from the right border skips the left border and the entire binary plasmid is transferred to the plant genome. In both these processes, the vector backbone with bacterial antibiotic resistance markers is transferred into the plant genome. Which is a potential biohazard. Transgenic plants with such transformation events should be carefully identified and eliminated.

A simple modification in the binary vector was carried out to address the problem of 'long transfer' and to enrich typical transformation events from right to left T-DNA borders. A CaMV35SP-*barnase*-INT cassette was placed to the left of the left T-DNA border as a non-T-DNA lethal gene. Any long transfer event beyond the left T-DNA border will lead to the expression of *barnase* resulting in them death and elimination such transformed cells. By deploying this strategy, the chances of the transfer of binary vector backbone were totally eliminated.

Biolistic Gene Transfer

For transforming plants, physical DNA delivery methods are useful. Such methods are microinjection, PEG-mediated transformation of protoplasts, electroporation of protoplasts and tissues and microprojectile bombardment of tissues. Among these, the microprojectile bombardment is the most widely deployed method for genotype-independent plant transformation, first used to deliver DNA and RNA into the epidermal cells of *Allium cepa*. Since then, it has been used to transform yeast and filamentous fungi, algae, cereals and pulses. Chloroplasts of *Chlamydomonas* and mitochondria of yeast and *Chlamydomonas* have also been biolistically transformed.

Particle bombardment use three systems for with various modifications. Electric discharge particle acceleration device ACCE L utilises an instrument to accelerate DNA-coated gold particles with any desired velocity by varying the input voltage. The biolistic PDS 1000 He device is powered by a burst of helium gas that accelerates the macrocarrier, DNA-coated gold particles are uniformly bombarded over target cells upon the macrocarrier. The advantages of particle bombardment system are: (1) Plants not infected by *Agrobacterium* can be transformed. (2) DNA may be transferred without using specialised vectors. (3) Multiple DNA fragments/plasmids can be introduced by co-bombardment, thus eliminating the need to construct a single large plasmid containing multiple transforming sequences (4) False positives resulting from reporter gene expression in *Agrobacterium* are avoided. (5) Transformation protocols are applicable to plants which lack good regeneration systems. (6) Organelle transformation is achieved only by particle bombardment.

The biolistic gene delivery system has been successfully used to create transgenics, but certain drawbacks of the technique, high copy number and rearrangements of transgene(s), have been observed thus causing gene silencing or genomic rearrangements. However, many recent improvements have been made to offset problems relating to gene silencing and genetic integrity. The overall transformation efficiency of biolistics has been improved by osmotic conditioning of cells.

Gene Silencing in Transgenic Plants

One of the common problems in making transgenic plants with increased level of endogenous gene expression by the use of heterologous strong promoters or a foreign gene of some utility, is silencing of transgenes in a vast population of transgenic plants at the transcription level. This is referred to as transcriptional gene silencing (TGS); whereas silencing at post-transcriptional level is referred to as post-transcriptional gene silencing (PTGS). TGS involves restriction of transcription and association with methylation of promoter region whereas in PTGS, though the genes are transcribed, their mRNA is degraded. PTGS is associated with methylation of the coding region of the transgenes.

TGS is associated with multiple copies of transgenes and inverted repeats (IRs) in transgenes, which can act in *cis* as well as in *trans*, indicating a homology-dependent mechanism of gene silencing. Plants may have also evolved TGS as a device to safeguard themselves from the activities of endogenous transposable elements, which are exist in multiple copies. However, PTGS has been observed even when unrelated foreign genes have been transferred to plants. PTGS has been associated with high copy number of transgenes, strength of the promoter or the stability of the transcripts. Above the threshold level for accumulation of transcripts PTGS sets in. Viruses infect plants and accumulate their transcripts at very high level. PTGS may stop transcription of viral genes. Some plants use PTGS to recover from viral infection and to stay free from future infections. As organisms compete with each other for their survival, some plant viruses have evolved mechanisms to counter or suppress PTGS.

DNA methylation is involved in both TGS and PTGS, but it is not clear if this is the cause or the effect of gene silencing. Apart from the common sites of methylation CG and of CNG symmetric sequences, there are also some methylation at asymmetric sites. In order to determine the involvement of DNA methylation in gene silencing, a mutant form of 35S promoter that lacked all symmetrical methylation acceptor sites was synthesised and used to drive the *bar* gene. When resistant lines obtained were crossed with plants carrying the 271-silencer locus a significant reduction in the number of phosphinothricin (PPT) resistant progeny was observed.

Only asymmetric methylation acceptor sites could allow the silenced lines be methylated. When the 271-silencer locus was outcrossed, PPT resistance was restored. Other studies but using wild type 35S promoter showed that even after outcrossing of the silencer locus the silenced state was maintained, suggesting that for initiation of silencing, methylation at symmetric methylation sites is not required. But this is essential for maintenance of silencing through generations.

Mutation in *MET* gene, encoding a DNA methyltransferase, drastically reduces DNA methylation and results in the release of silencing at most loci;

however some loci still remain silenced, probably because of the presence of other functional methyltransferases. Interestingly, the consequence the mutation in *DDM1* locus which encodes a plant homologue of SW12/SNF2 protein, a component of yeast chromatin remodelling complex, also results in loss of methylation and release of silencing, suggesting that normal chromatin structure is required for maintenance of normal methylation status as well as gene silencing. Studies with mammalian systems indicate that a methylated DNA-binding protein MeCP2, which can attach to even a single methylated CpG, recruits a repressor complex which has histone deacetylase. In a similar situation existing in plants, mutation in histone deacetylase (HADC) gene should be able to ease silencing without affecting methylation status. Transgenic inhibition of HADC in *Arabidopsis* is linked to pleiotropic effects similar to those observed in case of *MET/ DDMI* mutation, without any significant decrease in DNA methylation.

Mutation of *MOM* locus, which stands for 'morpheus molecule', also causes release of silencing without affecting DNA methylation. This gene could be another component of chromatin remodelling complex. TGS and PTGS operate differently but both involve DNA methylation, though of different regions. It has been proved that expressed RNA can trigger methylation of coding region as well as PTGS.

Upon infection, the potato spindle tuber viroid, which has a small RNA genome and does not code for any protein, can direct methylation of its transgene, integrated into plant genome. Production of double-stranded RNA complementary to coding sequence from a single transcript, which can pair to itself, can induce PTGS of genes having homologous sequence. Such double-stranded RNA can also be produced *in vivo* from promoter sequences having inverted repeats or other complex structures by read through transcription from nearby promoters. Recent experimental support this possibility.

Aberrant RNAs that are synthesized during PTGS are used by RNA-dependent RNA polymerase as substrates and synthesizes antisense RNAs. The pairing of both sense and antisense RNA leads to formation of double-stranded RNAs, which are targeted for degradation. Experimental evidence in tomato for the presence of RNA-dependent RNA polymerase has been obtained. The degradation products of about 25 bases to the silenced analogous ransgene have been observed during PTGS.

Transgene silencing is an unpredictable phenomenon, and hence plants showing silencing of the desired phenotype are discarded while plants which do not show silencing are used. Attempts to overexpress endogenous genes have often resulted in silencing of transgenes as well as endogenous genes, a phenomenon known as co-suppression, leading to PTGS, and attempts to overexpress genes of utility have suffered.

Attempts to overcome silencing by addressing the presence of duplicated sequences, high copy number of foreign genes, methylation acceptor sites, strength of promoter with silencing, are not well documented. Scientists have adopted the common occurrence of multiple copies of transgenes in case of biolistic method of transformation which has made *Agrobacterium*-mediated method of transformation more popular. However, there are reports of silencing of even single copy genes. Some attempts have been made to avoid silencing of transgenes.

Matrix attachment region (MAR) has been used in transformation vectors to prevent the influence of heterochromatin on the integrated genes. Insulation elements have proved beneficial in vertebrate systems to protect transgenes from effects of nearby *cis*-acting elements. Plant viruses have founds ways to avoid PTGS that plants use to avert viral infections. HC-protease (HC Pro) of potato virus Y (PVY) and 2b protein of cucumber mosaic virus (CMV) have been found to interfere with PTGS in tobacco plants infected by these viruses while HC Pro blocks the maintenance of gene silencing in tissues where silencing has already occurred, 2b protein prevents initiation of gene silencing at growing nodes of plants. Recently, a calmodulin-like protein, named rgs-CaM and identified from tobacco, has the ability to suppress PTGS and can be exploited to prevent gene silencing. However, whether such proteins would render these plants vulnerable to viral infections is yet to be investigated.

Depending upon the correlation of various features of vectors and gene silencing observed attempts should be made to avoid the use of potential methylation-acceptor sites. DNA repeats can be avoided by using different promoters and termination signals to drive expression of different genes of the vector. Vectors can be designed to incorporate MARs or insulators and even the genes for suppression of gene silencing could be incorporated.

Integration of Emerging Technologies of Transformation

Studying the developments in transformation, it is interesting to note that all improvements made in transformation systems were readily exploited for rice transformation. In the early stages, electroporation of protoplasts and PEG-mediated transformation of protoplasts were used to generate fertile transgenic rice plants. An electric discharge particle acceleration device was used to bombard immature rice embryos and generate fertile transgenic rice plants. Until 1994, it was strongly believed that rice is not amenable to *Agrobacterium*-mediated transformation. But with the construction and deployment of the superbinary vector pTOK233 for rice transformation, *Agrobacterium*-mediated rice transformation has become routine in many laboratories.

Agrobacterium-mediated cotransformation, a very simple and elegant approach for marker elimination, was simultaneously assessed in tobacco and rice. The cotransformation strategy was successfully utilised in rice for engineering provitamin A biosynthesis. A total of four genes contained in two separate T-DNA cassettes were transferred together.

Using tobacco MAR sequences, the usefulness of MARs in compartmentalising the expression of integrated transgenes was evident in the 140-fold increase in *gus* expression in rice. Immediately, a *hpt*-intron gene was successfully evaluated for rice transformation. As *gfp* was proving to be a useful, vital reporter of plants, a synthetic *gfp* version combined with *rbcS* promoter. Transit peptide was used for the first time to target the GFP protein to plastids to overcome its tendency to form insoluble complexes in cytoplasm. Isolated rice shoot apices were successfully used for *Agrobacterium mediated* transformation shoot apices were becoming attractive transformation targets. When an *ipt*-MAT strategy coupled to R/RS-based excision was used to obtain marker-free transgenic rice plants in one step, as demonstrated in rice.

Transformation technologies in rice are proving to be beneficial to meet the requirements of functional genomics. The clone with the *Xa21* gene was identified for bacterial blight resistance using a shot-gun approach with many rice genomic fragments. The clone new genes and to identify their functions, similarly generating 22,090 independent transgenic rice plants using *Agrobacterium*. It is estimated that as many as 25,700 TDNA tagged loci will be available for functional genomics analysis. Thus, with corresponding developments in transformation technologies and genome analysis, rice has become the first crop plant targeted for complete functional genomics analysis.

Direct Gene Transfer Methods

In 1984 the first transgenic plant was produced via Agrobacterium mediated modified transformation of *Nicotiana tabacum* protoplasts and thereafter several dozen plant species have been genetically engineered using different techniques.

Other techniques such as selectable markers were also developed that facilitated the development in genetic engineering for obtaining transformed plants. Since this technique is not suitable for monocotyledon plants as they are not natural host of Agrobacterium, other methods of direct gene transfer have been developed for use with monocots and other species, be categorised on the basis of the use of protoplasts or cell and tissue as the target materials. Freshly isolated protoplasts are used for genetic transformation using direct gene transfer methods.

a) *DNA Transfer in protoplasts:* This is accomplished by:
   i) Electroporation.
   ii) Chemically stimulated DNA uptake by protoplasts
   iii) Liposomes
   iv) Microinjection
   v) Sonication

b) *DNA Transfer in Plant Tissues:*
   i) Acceleration of DNA coated microparticles
   ii. Laser microbeam
   iii. Silicon carbide fibres

Direct uptake of DNA by isolated protoplasts is a genotype-dependent response, and regeneration from protoplasts is not common in all the species. Hence production of fertile transgenic plants has posed problems in most of the cereal species. However, transgenic fertile plants have been produced in *Sorghum vulgare, Oryza sativa* and *Hordeum vulgare*. Direct delivery of free DNA molecules into plant protoplasts by physical (electroporation and micro injection) and chemical (polyethylene glycol) methods have been developed to facilitate DNA delivery across the plasma membrane.

DNA Transfer in Protoplasts

Electroporation

Electroporation is based on the use of the short electrical pulses of high field strength, causing the uptake of DNA into protoplasts by temporary permeabilisation of the plasma membrane to macromolecules. Protoplasts and foreign DNA, placed in a buffer between two electrodes are treated to a high intensity electric current which damages membranes and creates pores in membranes. Immediately thereafter DNA diffuses through these pores, until the pores are resealed. The technique is optimised by using appropriate electric field strength. The optimum field strength that is dependent on the following:

i) The pulse length of electric current
ii) Composition and temperature of the buffer solution
iii) Concentration of foreign DNA in the suspension
iv) Protoplasts density, and
v) Size of the protoplasts.

The removal of pectin from the plant wall increases the amount of DNA which can be introduced by electroporation, tobacco mosaic virus being a case point. Electroporation has been used successfully for transient and stable transformation of protoplasts from a wide range of species. Plating efficiency of electroporated protoplasts grown on selection medium can be as high as 0.5 per cent, for tobacco being 0.2 per cent of electroporated leaf mesophyll protoplasts giving rise to transgenic calli. Low transformation efficiency is common in cereals, e.g. in rice 0.002 efficiency was recorded.

Chemically stimulated DNA uptake

Direct uptake of DNA by protoplasts is stimulated by polyethylene glycol (PEG) which is the most widely used chemical for this purpose. Transformation involves stimulated by PEG mixing of freshly isolated protoplasts with DNA and immediately adding 15-20 per cent PEG dissolved in a buffer containing divalent cations. This mixture is incubated for 30 minutes, protoplasts are washed and then plated in petri plates for culture and growth. The optimisation of transformation frequencies by this method include factors that follows.

i) PEG concentration in the mixture
ii) Composition and concentration of salts used
iii) The pH of the solution
iv) Concentration of the foreign DNA

v) Size and form (linear, supercoiled) of the DNA molecules used

vi) Culture and selection techniques used for protoplasts.

There is preference for PEG mediated transformation over electroporation for stable transformation of monocot protoplasts due to relatively higher survival rates after treatment.

While PEG also stimulates the uptake of liposomes, improves the efficiency of electroporation, and causes precipitation of ionic macromolecules like DNA, it also stimulates their uptake by endocytosis. PEG mediated DNA uptake typically transforms 0.1 to 0.4 per cent of the total protoplasts treated. Production of transgenic plants depends upon the regeneration competence of the transformed protoplasts. In Petunia, 40 per cent transformed calli derived from mesophyll protoplasts could be induced to form fertile plants, equal to about 0.1 per cent transformation efficiency of the treated protoplasts. In different species different transformation efficiency has been observed, e.g., in embryogenic protoplasts suspension of rice was 0.0004 per cent and, soyabean and tobacco was 0.7-1 per cent.

Liposomes

Liposomes, small lipid sacs containing plasmids and prepared artificially have also been used as a carrier for the introduction of nucleic acid into plant protoplasts. PEG stimulates the fusion of liposomes with plant protoplasts. Liposomes mediated transformation includes positively charged agents such as cations in the transformation mixture or using the cationic liposome preparation.

For maize protoplast, other chemical agents like polycation polybrene or lipofectin have also been used for both transient and stable transformation. The new protoplasts transformation methods are cationic liposome and polycation mediated DNA delivery which are considered better than other methods of transformation. There are several advantages for the use of this technique.

i) Protection of DNA/RNA from nuclease digestion.

ii) They have low cell toxicity.

iii) Encapsulation of nucleic acids makes them more stable during storage.

iv) High degree of reproducibility.

v) This method is applicable to wide range of plant cell types.

Microinjection

Microinjecting nucleic acids into protoplasts or intact cells via is a labour-intensive procedure requiring special capillary needles, pumps, micromanipulators, inverted microscope and other equipment. However, nucleus or cytoplasm can be injected and cells can be cultured individually to produce callus or plants, thus avoiding transformants by drug resistance or marker genes. The worker has to skilfully insert the needle into the cytoplasm or in the nucleus. The basic technique is similar to that used for animal cell microinjection. In order to microinject protoplasts or other plant cells, the cells need to be immobilised which are done by:

i) The use of a holding pipette which holds the cells by vacuum.

ii) Attachment of cells to poly-L-lysine coated cover slips.

iii) Embedding the cells in agarose, agar or sodium alginate.

Glass micropipettes are with openings of about 0.3 uM in diameter, are inserted into plant cell cytoplasm and nuclei with the aid of a micro manipulator device. A syringe-like device is used for the controlled delivery of volume (10-11 - 10-4 ul) into the plant cell. Most plant cells are injected while keeping inside microdroplets (2-50 ul) of medium using a chamber which is sterile, vibration free and permits temperature and humidity regulation. Using this procedure, a maximum of 100-200 cells per hour can be microinjected.

The regeneration ability of the microinjected cells decides the recovery of transformants. Different methods have been used to grow injured single cells or protoplasts. Hanging droplets, covered under a thin layer of agar or agarose, and micro culture are being used. Attempts have been made to inject linear, or super coiled DNA, in cytoplasm or in nucleus. Nuclear injections are found better for transformations.

Sonication

In protoplasts of sugar beet and tobacco mild sonication facilitates the uptake and transient expression of a chloramphenicol acetyltransferase (CAT) gene. This is a better method than electroporation method used for the same material. Plating efficiency was also similar to untreated cells. However, transgenic plant production using this technique has not been reported so far.

Free DNA Transfer to Intact Tissue

Acceleration of DNA Loaded Microparticles

The latest technology to transfer DNA into intact tissues involves the transfer of micro-sized particles coated with DNA to penetrate the cells. Here micron size tungsten or gold particles are accelerated in a gun barrel to velocities sufficient for non-lethal penetration of cell walls and membranes. In 1987 the particle gun was first used to transfer chimeric DNA and viral RNA molecules into intact onion cells. Tungsten acted as a carrier of nucleic acids as it was available in micro-size balls, non-toxic to cells and was dense enough for rapid penetration of target material. Microprojectile mediated transformation, is a mechanical method of introducing DNA in to any plant species, can be successfully used where plasmids or protoplasts mediated transformation cannot be used. An acceleration device used to propel particles carrying plasmid DNA based on machine or technique used to accelerate the particles, are called 'particle gun technology', 'biolistic method', 'DNA bombardment', 'particle acceleration of DNA method' and 'electric discharge particle acceleration method'.

This technique of stable transformation and testing a gene for cell and organelle specific expression has three components:

i) The basic equipment to generate particle acceleration.

ii) Metal particles coated with precipitated DNA (desired gene).

iii) Plant tissues to be used for particle penetration. The method for regeneration should be previously standardised and proper tissues be selected for bombardment.

In an electric discharge particle acceleration device a high voltage discharge (14 KV current) is delivered to a small water droplet which quickly vaporises and releases energy to propel DNA coated gold spheres into target cells. Similarly, a DNA carrier is attracted due to potential differences, stopped in-between by a screen, DNA coated particles cross the screen and fly towards target tissue and deliver the DNA into cells.

DNA Coating

This sophisticated technology requires precise preparation of DNA coated gold or tungsten particles, having the following properties:

i) High density (19 g/cm^3 or greater) to ensure proper acceleration and penetration through cell walls.

Size (0.5-5 um) should match with size of the cells. Large sized spheres can be used with large cell size.

Gold is costlier than tungsten, but it does not oxidise like tungsten.

DNA is precipitated onto the particles prior to bombardment. Most commonly, $CaCl_2$ and spermidine are used to precipitate plasmid DNA onto tungsten particles. Ethanol is also used for precipitation of the DNA on gold particles.

Plant Tissue

Competent generating plant tissue should be used for transformation. Though mostly embryogenic tissue is an ideal material for transformation, yet leaf bombardment in tobacco, and stem bombardment in cranberry have resulted in transformed plants. Normally, reporter gene and selectable marker genes are used to isolate and select transformed cells/plantlets.

Laser Microbeam

Laser beam has been usefully employed for transformation of plant cells. To introduce DNA into plant cells and chloroplasts an ultraviolet (UV) laser. 343 nm beam is directed through an adjustable attenuator into the optical path of an inverted microscope. The focus of the laser beam is adjusted so that its focus is identical with that of the objective lens on a specimen in the microscope. This laser beam can then puncture holes in any part of cell which is in focus. Laser micropuncture of the cell wall and plasma membrane allows uptake of plasmid DNA into cells using Brassica napus cells and microspores. This technique has also been used to transfer genes into isolated chloroplasts and chloroplasts of intact protoplasts resulting in 20 per cent transformation achieved by this method but fertile plants are yet to be produced by this method.

References

Atherton, K. T., ed., *Genetically Modified Crops: Assessing Safety*, London and New York: Taylor and Francis, 2002.

British Medical Association. *The Impact of Genetic Modification on Agriculture, Food and Health*. BMJ Books. 1999.

Debergh Zimmerman, (ed.) *Micropropagation: technology and application*. Kluwer Academic Publishers. 1991.

Villalobos, V.M., Ferreira, P. & Mora, A. "The use of biotechnology in the conservation of tropical germplasm". *Biotech. Adv*., 9: pp 197-215. 1991.

4

Transgenic Plants with Beneficial Traits

During the last decades, a tremendous progress has been made in the development of transgenic plants using the various techniques of genetic engineering. The plants, in which a functional foreign gene has been incorporated by any biotechnological methods that generally are not present in the plant, are called transgenic plants. As per estimates recorded in 2002, transgenic crops are cultivated world-wide on about 148 million acres (587 million hectares) land by about 5.5 million farmers. Transgenic plants have many beneficial traits like insect resistance, herbicide tolerance, delayed fruit ripening, improved oil quality, weed control etc.

Some of the commercially grown transgenic plants in developed countries are: "Roundup Ready" soybean, 'Freedom II squash', 'High-lauric' rapeseed (canola), 'Flavr Savr' and 'Endless Summer' tomatoes. During 1995, full registration was granted to genetically engineered Bt gene containing insect resistant 'New Leaf' (potato), 'Maximizer' (corn), 'BollGard' (cotton) in USA. Some of the traits introduced in these transgenic plants are as follows:

Traits Introduced in Transgenic Plants

Stress Tolerance

Biotechnology strategies are being developed to overcome problems caused due to biotic stresses (viral, bacterial infections, pests and weeds) and abiotic stresses (physical actors such as temperature, humidity, salinity etc).

Abiotic Stress Tolerance

The plants show their abiotic stress response reactions by the production of stress related osmolytes like sugars (e.g. trehalose and fructans), sugar alcohols (e.g. mannitol), amino acids (e.g. proline, glycine, betaine) and certain proteins (e.g. antifreeze proteins). Transgenic plants have been produced which over express the genes for one or more of the above mentioned compounds. Such plants show increased tolerance to environmental stresses. Resistance to abiotic stresses includes stress induced by herbicides, temperature (heat, chilling, freezing), drought, salinity, ozone and intense light. These environmental stresses result in the destruction, deterioration of crop plants which leads to low crop productivity. Several strategies have been used and developed to build ressitance in the plants against these stresses.

Herbicide Tolerance

Weeds are unwanted plants which decrease the crop yields and by competing with crop plants for light, water and nutrients. Several biotechnological strategies for weed control are being used e.g. the over-production of herbicide target enzyme (usually in the chloroplast) in the plant which makes the plant insensitive to the herbicide. This is done by the introduction of a modified gene that encodes for a resistant form of the enzyme targeted by the herbicide in weeds and crop plants. Roundup Ready crop plants tolerant to herbicide-Roundup, is already being used commercially.

The biological manipulations using genetic engineering to develop herbicide resistant plants are:

(a) over-expression of the target protein by integrating multiple copies of the gene or by using a strong promoter.,

(b) enhancing the plant detoxification system which helps in reducing the effect of herbicide.,

(c) detoxifying the herbicide by using a foreign gene., and

(d) modification of the target protein by mutation.

Some of the examples are:

1. *Glyphosate resistance.* Glyphosate is a glycine derivative and is a herbicide which is found to be effective against the 76 of the world's worst 78 weeds. It kills the plant by being the competitive inhibitor of the enzyme 5-enoyl-pyruvylshikimate 3- phosphate synthase (EPSPS)

in the shikimic acid pathway. Due to it's structural similarity with the substrate phosphoenol pyruvate, glyphosate binds more tightly with EPSPS and thus blocks the shikimic acid pathway.

Certain strategies were used to provide glyphosate resistance to plants.

(a) It was found that EPSPS gene was overexpressed in Petunia due to gene amplification. EPSPS gene was isolated from Petunia and introduced in to the other plants. These plants could tolerate glyphosate at a dose of 2- 4 times higher than that required to kill wild type plants.

(b) By using mutant EPSPS genes- A single base substitution from C to T resulted in the change of an amino acid from proline to serine in EPSPS. The modified enzyme cannot bind to glyphosate and thus provides resistance.

(c) The detoxification of glyphosate by introducing the gene (isolated from soil organism- Ochrobactrum anthropi) encoding for glyphosate oxidase into crop plants. The enzyme glyphosate oxidase converts glyphosate to glyoxylate and aminomethyl-phosponic acid. The transgenic plants exhibited very good glyphosate ressitance in the field.

2. *Phosphinothricin resistance.* Phosphinothricin is a broad spectrum herbicide and is effective against broad-leafed weeds. It acts as a competitive inhibitor of the enzyme glutamine synthase which results in the inhibition of the enzyme glutamine synthase and accumulation of ammonia and finally the death of the plant. The disturbace in the glutamine synthesis also inhibits the photosynthetic activity.

   The enzyme phosphinothricin acetyl transferase (which was first observed in Streptomyces sp in natural detoxifying mechanism against phosphinothricin) acetylates phosphinothricin, and thus inactivates the herbicide. The gene encoding for phosphinothricin acetyl transferase (bar gene) was introduced in transgenic maize and oil seed rape to provide resistance against phosphinothricin.

Other Abiotic Stresses

The abiotic stresses due to temperature, drought, and salinity are collectively also known as water deficit stresses. The plants produce osmolytes or osmoprotectants to overcome the osmotic stress. The attempts are on to use genetic engineering strategies to increase the production of osmoprotectants in the plants. The biosynthetic pathways for the production of many osmoprotectants have been established and genes coding the key enzymes have been isolated. E.g. Glycine betaine is a cellular osmolyte which is produced by the participation of a number of key enzymes like choline dehydrogenase, choline monooxygenase etc. The choline oxidase gene from Arthrobacter sp. was used to produce transgenic rice with high levels of glycine betaine giving tolerance against water deficit stress.

Scientists also developed cold-tolerant genes (around 20) in Arabidopsis when this plant was gradually exposed to slowly declining temperature. By introducing the coordinating gene (it encodes a protein which acts as transcription factor for regulating the expression of cold tolerant genes), expression of cold tolerant genes was triggered giving protection to the plants against the cold temperatures.

Insect Resistance

A variety of insects, mites and nematodes significantly reduce the yield and quality of the crop plants. The conventional method is to use synthetic pesticides, which also have severe effects on human health and environment. The transgenic technology uses an innovative and eco-friendly method to improve pest control management.About 40 genes obtained from microorganisms of higher plants and animals have been used to provide insect resistance in crop plants

The first genes available for genetic engineering of crop plants for pest resistance were Cry genes (popularly known as Bt genes) from a *bacterium Bacillus thuringiensis*. These are specific to particular group of insect pests, and are not harmful to other useful insects like butter flies and silk worms. Transgenic crops with Bt genes (e.g. cotton, rice, maize, potato, tomato, brinjal, cauliflower, cabbage, etc.) have been developed. This has proved to be an effective way of controlling the insect pests and has reduced the pesticide use. The most notable example is Bt cotton (which contains CrylAc gene) that is resistant to a notorious insect pest Bollworm *(Helicoperpa armigera)*.. There are certain other insect resistant genes from other microorganisms which have been used for this purpose. Isopentenyl transferase gene from Agrobacterium tumefaciens has been introduced into tobacco and tomato. The transenic plants with this transgene were found to reduce the leaf consumption by tobacco hornworm and decrease the survival of peach potato aphid.

Certain genes from higher plants were also found to result in the synthesis of products possessing insecticidal activity. One of the examples is the Cowpea trypsin inhibitor gene (CpTi) which was introduced into tobacco, potato, and oilseed rape for develping transgenic plants. Earlier it was observed that the wild species of cowpea plants growing in Africa were resistant to attack by a wide range of insects. It was observed that the insecticidal protein was a trypsin inhibitor that was capable of destroying insects belonging to the orders Lepidoptera, Orthaptera etc. Cowpea trypsin inhibitor (CpTi) has no effect on mammalian trypsin, hence it is non-toxic to mammals.

Virus Resistance

There are several strategies for engineering plants for viral resistance, and these utilizes the genes from virus itself (e.g. the viral coat protein gene). The virus-derived resistance has given promising results in a number of crop plants such as tobacco, tomato, potato, alfalfa, and papaya. The induction of virus resistance is done by employing virus-encoded genes-virus coat proteins, movement proteins, transmission proteins, satellite RNa, antisense RNAs, and ribozymes. The virus coat protein-mediated approach is the most successful one to provide virus resistance to plants. It was in 1986, transgenic tobacco plants expressing tobacco mosaic virus (TMV) coat protein gene were first developed. These plants exhibited high levels of resistance to TMV.

The transgenic plant providing coat protein-mediated resistance to virus are rice, potato, peanut, sugar beet, alfalfa etc. The viruses that have been used include alfalfa mosaic virus (AIMV), cucumber mosaic virus (CMV), potato virus X (PVX), potato virus Y (PVY) etc.

Resistance Against Fungal and Bacterial Infections

As a defense strategy against the invading pathogens (fungi and bacteria) the plants accumulate low molecular weight proteins which are collectively known as pathogenesis-related (PR) proteins.

Several transgenic crop plants with increased resistance to fungal pathogens are being raised with genes coding for the different compounds. One of the examples is the Glucanase enzyme that degrades the cell wall of many fungi. The most widely used glucanase is beta-1,4-glucanase. The gene encoding for beta-1,4 glucanase has been isolated from barley, introduced, and expressed in transgenic tobacco plants. This gene provided good protection against soil-borne fungal pathogen Rhizoctonia solani.

Lysozyme degrades chitin and peptidoglycan of cell wall, and in this way fungal infection can be reduced. Transgenic potato plants with lysozyme gene providing resistance to Eswinia carotovora have been developed.

Delayed Fruit Ripening

The gas hormone, ethylene regulates the ripening of fruits, therefore, ripening can be slowed down by blocking or reducing ethylene production. This can be achieved by introducing ethylene forming gene(s) in a way that will suppress its own expression in the crop plant. Such fruits ripen very slowly (however, they can be ripen by ethylene application) and this helps in exporting the fruits to longer distances without spoilage due to longer-shelf life.

The most common example is the 'Flavr Savr' transgenic tomatoes, which were commercialized in U.S.A in 1994. The main strategy used was the antisense RNA approach. In the normal tomato plant, the PG gene (for the enzyme polygalacturonase) encodes a normal mRNA that produces the enzyme polygalacturonase which is involved in the fruit ripening. The complimentary DNA of PG encodes for antisense mRNA, which is complimentary to normal (sense) mRNA. The hybridization between the sense and antisnse mRNAs renders the sense mRNA ineffective. Consequently, polygalacturonase is not produced causing delay in the fruit ripening. Similarly strategies have been developed to block the ethylene biosynthesis thereby reducing the fruit ripening. E.g. transgenic plants with antisense gene of ACC oxidase (an enzyme involved in the biosynthetic process of ethylene) have been developed. In these plants, production of ethylene was reduced by about 97% with a significant delay in the fruit ripening.

The bacterial gene encoding ACC deaminase (an enzyme that acts on ACC and removes amino group) has been transferred and expressed in tomato plants which showed 90% inhibition in the ethylene biosynthesis.

Male Sterility

The plants may inherit male sterility either from the nucleus or cytoplasm. It is possible to introduce male sterility through genetic manipulations while the female plants maintain fertility. In tobacco plants, these are created by introducing a gene coding for an enzyme (barnase, which is a RNA hydrolyzing enzyme) that inhibits pollen formation. This gene is expressed specifically in the tapetal cells of anther using tapetal specific promoter TA29 to restrict its activity only to the cells involved in pollen production. The restoration of male fertility is done by introducing another gene barstar that suppresses the activity of barnase at the onset of the breeding season. By using this approach, transgenic plants of tobacco, cauliflower, cotton, tomato, corn, lettuce etc. with male sterility have been developed.

Reversible Male Sterility in Plants

Heterozygous individuals being often healthier and stronger than homozygous ones, the only way to guarantee heterozygoiscity in plants is to make sure self-pollination cannot occur. Earlier it was difficult for most crop plants to exclude selfing. To exclude self-pollination, it would be good to introduce male sterility in plants: progeny from such plants are then expected to be 100 per cent heterozygous. A promoter was identified that was turned on exclusively in tapetum cells to introduce male sterility. This promoter then was linked up to a gene coding for a bacterial ribonuclease which selectively chops up ribonucleic acids. The promoter/ribonuclease construct was then generated into plants. Because the promoter allows expression only in tapetum cells, the gene construct disrupts only growth of the tapetal tissue and its end product, pollen. Plants transformed with this construct were male-sterile but otherwise normal.

Male-sterile plants are valued for hybrid seed production, but their value is limited when it comes to crop production. Crops, such as wheat, rice and tomato, in which the seed or fruit is the harvested product, must have the fertility restored. Fortunately, since the ribonuclease is inhibited very much by a simple protein, named barstar, the male-sterile plant can be crossed with a male-fertile variety in which the gene for barstar has been introduced, and the result is progeny with viable pollen and restored fertility.

Antisense RNA

Antisense RNA are nucleotide strands that are grown in a cell and that are complementary to a particular mRNA. Antisense RNA can be produced, for example, by inverting the coding region of a gene with respect to its promoter, and hybridised with its corresponding mRNA, making it double-stranded. The double-stranded mRNA no longer can be recognised by the protein-synthesizing machinery, and thus expression of this mRNA is suppressed or broken down quickly. Thus, one can inactivate specific genes while not interfering with others.

Antisense approaches protect plants from damage by plant viruses. For example, reversal of a gene from bean yellow mosaic virus (BYMV), and putting it into tobacco under a reasonably strong promoter, has resulted in a BYMV-resistant tobacco variety. A similar approach is used to transfer viral resistance to other plants which is significant, in that currently no effective, environmentally friendly methods exist to control many plant viruses.

The RNAi approach is very useful for both agriculture and medicine. The technology provides an excellent approach for reverse genetics in eukaryotes. With this method one can turn off genes, and see what the consequences are.

Agricultural Applications

Two main reasons for the more rapid commercialisation of bioengineered crops in the developing world are: (a) less tight governmental approval mechanisms, and (b) hungrier populations. In developed countries biotechnological applications may be utilised to reduce production costs; in contrast, in the developing world the main factor is the production of more food. Indeed, genetic engineering applications seem to be pretty successful to cut down on pathogen-induced losses. For example, genetic modification of papaya plants protects very well against the very destructive ringspot virus.

Herbicide Resistance

Weeds compete with crops for water, nutrients and sunlight and can decrease crop yields from 35 per cent to 100 per cent. A number of options for minimising the impact of weeds on crop productivity are available to growers; one option is application of herbicides to the weeds.

Normally all crops have some degree of innate resistance to some herbicides. Today agriculturists minimise the negative impact of weeds by using herbicides and herbicide resistant varieties of most crops, including all of the major commodity crops, as well as small acreage horticultural crops, such as vegetables. An attendant benefit of herbicides is the impact their use has on soil erosion. Herbicides and herbicide- resistant crops allow growers to use minimal tillage or no-till techniques, which significantly reduces the impact of agriculture on soil erosion. Today more than 290 types of herbicide resistant weeds are present in agricultural fields around the world.

More than half a century ago, most herbicides used in agriculture were developed by generating many compounds in the lab and spraying them on weeds in the greenhouse to identify likely candidates for further development. Those that were herbicidal were then tested on crops to assess the crops' sensitivity. On obtaining promising candidates for new herbicidal compounds, tests were then conducted to determine whether the herbicide and herbicide resistant crop performed well under field conditions and to assess any impacts on the environment and human health.

The number of chemicals to be tested to develop a new herbicide began to increase exponentially from less than 1,000 in 1950 to approximately one million today. Consequently companies began to focus on developing new weed management options by modifying the crop, rather than finding a new herbicide. Beginning with herbicide already proven to be effective and then giving the crop plant the capacity to resist that herbicide is more efficient and allows herbicide companies to expand the applicability of their most effective and safest products.

Prior to the advent of modern biotechnology crop developers had to perforce use plants in the same or very closely related species, and most crop plants have very little genetic diversity for herbicide resistance. An exception was endogenous tolerance of soyabeans to metribuzin and 2,4-D, which varied among different cultivars. As a result, soyabean breeders were able to use conventional breeding and selection to increase soyabean's endogenous tolerance to these two herbicides, and to combine resistance to both in a cultivar that was resistant to one, but not the other. Crop developers have employed a series of techniques to create novel herbicide resistance traits. Screening and selection of plants with novel genetic variability led to the development of soyabeans tolerant to sulfonylurea herbicides.

Recent successes in plant cell and tissue culture linked with modern biotechnology enable researchers to screen cells and undifferentiated plant callus for herbicide resistance, which increases development efficiency. Treatment of corn tissue cultures with imidazolines followed by recurrent selection created imidazoline- resistant callus that could then be regenerated into corn plants and bred into a variety of cultivars.

Crop developers have also exploited the capacity of weeds to evolve resistance. For example, a breeding programme was initiated to move triazine resistance found in a weed, *Brassica rapa,* to canola, a crop closely related to the weed. Triazine-resistant canola was created, but unfortunately, the new resistance trait reduced crop productivity, which affected the cultivar's commercial value.

The tools of modern biotechnology have made it possible for a herbicide resistant gene present in a given organism to be incorporated into virtually any crop. This flexibility has revolutionised weed management by expanding the spectrum of herbicides that a crop can tolerate.

The first transgenic herbicide resistant crop, bromoxynil-resistant cotton, was commercialised in 1995, and glyphosate-resistant soyabeans came to market the following year. Glyphosate resistant soybeans, corn, cotton and canola are now grown on millions of hectares around the world.

Impact of Herbicide-Resistant Biotech Crops

The new herbicide resistant crops are a great boon to growers as the easy to use, low-cost and effective weed control programme does not harm their crops. Notwithstanding, the new herbicide resistant crops, developed with modern biotechnology, have significantly added to the growers' profits by decreasing their input costs, increasing yields, or both, in some cases.

These benefits have enhanced, the rate of adoption of biotech herbicide-resistant crops. Glyphosate-resistant soyabeans 60 per cent of the global soyabean acres. In 2005, 71 per cent (157 million acres) of the 222 million acres of biotech crops planted globally were herbicide-resistant varieties of corn, soyabean, canola and cotton. The majority of the corn, canola and cotton varieties were resistant to glyphosate and as a result, glyphosate-resistant weeds have begun appearing in fields with row crops like corn and soyabean where glyphosate has been used repeatedly. The effectiveness of glyphosate-resistant weed control deterred farmers from implementing diverse weed management practices, and lack of diversity in weed control measures encourages the evolution of resistance to a measure that is used repeatedly.

Future Prospects

All glyphosate-resistant biotech crops worldwide depend mainly on providing crops with a gene that encodes a glyphosate-insensitive form of the glyphosate target protein. Recently scientists have discovered novel sources of resistance to the sulfonylureas and imidazoline herbicides as well as glyphosate. A new biochemical mechanism of glyphosate resistance was recently discovered in a naturally-occurring microorganism which has a gene for an enzyme capable of deactivating glyphosate. Scientists optimised the activity of this enzyme. Once they transfer the gene for the optimised enzyme into crops, it will effectively protect the crops from glyphosate treatment.

Scientists have also found a new source of resistance to sulfonylureas and imidazolines in corn and soyabeans. When this mechanism is adequately expressed, the sulfonylureas, imidazolines and a number of other similar herbicides can be applied without harming the crop. These herbicides, like glyphosate, are highly advantageous to growers because they are effective against a broad spectrum of weeds.

These combinations of traits in one crop will provide growers with huge flexibility in weed control and a new strategy for controlling their most difficult weed problems.

Insect Resistance

Since development of insect-resistant plants in 1782 numerous insect-resistant plants have been developed by research centres through conventional method or biotechnology. For example, the International Rice Research Institute (IRRI) has developed and released numerous rice varieties resistant to brown planthopper, green leafhopper, rice stem borer, and rice gallmidge. Several transgenic insecticidal cultivars obtained through biotechnology approach have also been developed, such as Bt-corn, Bt-cotton, and Bt-rice.

Farmers all over the world are using many insect-resistant plants. Rice resistance to the rice brown planthopper and green rice leafhopper has been planted over 20 million hectares in Asia. Wheat resistant to Hessian fly are grown in the mid-western United States. In 1999, Bt-plants resistance to insect was planted over 8.9 million hectares, mostly in United Stated. The wide spreaduse of insect-resistant plants may be because these plants are the most economic, least complicated and environmentally soundest approach for protecting plants from insect damage.

New insect biotypes, such as brown planthopper biotypes and Hessian fly biotypes, have been developed. There has been development of new insect biotypes overcoming transgenic insecticidal plants derived from biotechnology approach which has not been recorded so far. Two strategies to delay the development of new insect bio-types have been reported. The first strategy relates to insect-plant resistant development and the second, to their application in the field in which insect-resistant plants are one of the components of integrated pest management approach.

Method to Develop Insect-Resistant Plants

In breeding programme for insectresistance plants, are first identified the parents or donors of resistance. These may be cultivated germplasm, landraces, weeds races, wild species, or different species. The breeding method is based on the following factors: the source of donor of resistance, the efficiency and certainty of selection of progenies. If the donor of resistance is of commercial varieties, they may be improved by pure line or mass selection or hybridised with elite germplasm. If they are land-races, weed races, or wild species, they have to be hybridised with the elite germplasm. In the case of the donor of resistance being of different species, biotechnology method may used, such as genetic transformation or protoplast fusion. If the donors are not available, biotechnology method may be used, such as somaclonal variation for insect-resistance. If the efficiency and certainty in the selection of desired traits are needed, the use of biotechnology method, such as marker assisted selection and anther culture, may be appropriate.

The method to develop insect-resistant plant can be grouped into conventional and biotechnology. The conventional approaches use the whole living organism and includes pure line selection, mass selection, hybridisation and while biotechnology approaches use cells or biomolecules to develop insect-resistant plants and includes protoplast fusion, somaclonal variation, molecular assisted selection, and genetic transformation. For anther culture and embryo culture, earlier they used organism of F1 plants to produce pollen grain and wild and cultivated species to produce embryo, but now they use cells culture, like the anther culture and embryo culture as a biotechnology approach.

Anther Culture

Using an artificial medium, anther or pollen grain can be cultured *in vitro*. On the artificial medium anther may form callus, shoot, root, and finally the entire plants. All the plants are haploid. This approach is helpful in speeding up the formation of homozygous population of insect-resistant plants.

Embryo Culture

Wide hybridisation involves transfer of genes conferring insect-resistant from wild species, that are often more resistant to insects to cultivated plants, but this will produce abnormal interspecific hybrid embryos. The embryo can be saved by culturing it on the nutrient medium to generate the entire plants. Several genes for insect resistance have been transferred from wild to cultivated germplasm (Table 1).

Table 1. Genes conferring insect resistance transferred from wild to cultivated species

Recipient	*Alien donor*	*Trait*
Bread wheat	*Secale cereale*	Resistance to greenbug
	S. cereale	Resistance to Hessian fly
	Aegilops squarrosa	Resistance to Hessian fly
	Rice Oryza officinalis	Resistance to BPH
	O. officinalis	Resistance to WBPH
	O. australiensis	Resistance to BPH
	O. minuta	Resistance to BPH
Peanut	*Arachis monticola*	Resistance to chewing insect
Lettuce	*Lactusa virusa*	Resistance to Aphids
Cotton	*Gossypium armourianum*	Boll weevil, leafworm, bollworm

Protoplast Fusion

Protoplasts are plant cells that could be isolated by digested the wall enzimatically. The traits of resistance to the pest may be present in the one of two species that cannot be hybridised sexually. The two species may be formed from hybrids through protoplasts fusion. The protoplasts may be cultured on artificial medium and some protoplasts will grow into entire plants which will carry the resistant traits.

Somaclonal Variation

Selecting insect-resistant variants can be done by somaclonal variation. Insect-resistant somaclones can be selected through the following steps (a) calli or cell suspension derived from high-yielding variety grown for several or long-term cycles, (b) the long-term cell lines regenerated into plants, and

(c) the regenerated plants evaluated against target insects. About 2,000 plantlets of sugarcane were assessed for resistance to the sugarcane borer under artificial infestation and natural infestation in field plots. Some somaclones showed resistance to sugarcane borer.

Marker-assisted Selection

Nucleic acid probe, antibodies, or enzymes may be used as a marker-assisted selection in breeding programme to determine the genetic constitutions of plants or plant extracts, including the presence of resistant traits. The precise information from this method will increase the speed and certainty of selection progeny carrying resistance traits in conventional plant breeding.

Genetic Transformation

Genetic transformation integrates cloned DNA segments or genes from plants, bacteria, or animals into a new genetic background. These DNA segments or genes may confer resistance to insect. The genes may be delivered to the protoplast or cell by using one of the following method: Agrobacterium, biolistic, electroporation, PEG, vortex, and microinjection. Several resistant plants have been developed through genetic transformation (Table 2).

Table 2. Resistant plants develop through genetic transformation

Plants	*Traits*
Alfalfa	Insect resistance
Cabbage	Insect resistance
Cotton	Resistance to bollworms and budworm
Eggplant	Insect resistance
Maize/Corn	Resistance to corn borer
Potato	Resistance to Colorado potato beetle
	Resistance to Potato tuber moths
Soybean	Insect resistance
Sugar cane	Insect resistance
Sweet potato	Insect resistance
Tobacco	Insect resistance
Tomato	Insect resistance
Rice	Insect resistance (Bt toxin)

Strategies to Develop Insect-resistant Plants

To anticipate the breakdown of insect-resistant plants due to the development of new bio-types, various strategies should be adopted in developing those plants. One strategy is to prolong the useful life of insect-resistant plants by developing durable insect-resistant plants, governed by polygenes. Durable resistance is also referred to poligenic resistance or horizontal resistance. The other strategy is to develop varieties with different genes conferring insect-resistant which may be achieved by developing vertical resistant plants.

The vertical-resistant plant is governed by major genes. The more vertical resistant varieties with different genes is available, the easier it is for farmers to access new varieties to replace the previous varieties.

Durable Insect-resistant Plants

The level of resistance of durable insects-resistant plants is generally not very high the development of new biotypes is very slow. Polygenes resistance could be obtained from the landraces. Once crossing has been made between cultivated and landraces, there is the difficulty in selecting the desired segregants as not all the polygenes or QTL from landraces are transferred to cultivated, and the level of resistance is diluted. The drawback is that the screening techniques currently available is only applicable to identify segregants with high level of resistance, not segregants with low level of resistance.

In molecular-assisted selection, facilitating the development of durable insect-resistant plants, the first step should be to tag the QTL for insect resistance with molecular markers. Molecular marker-based selection will assist in the accumulation of polygenes. QTLs for resistance to rice brown planthopper have been tagged with RFLP markers.

Vertical Insect-resistant Plants

Vertical insect-resistant plants are more easier to develop since major genes are more easier to transfer from one variety to others. Numerous vertical insect-resistant plants with different genes should be developed through conventional or biotechnology in order to expect the new insect biotypes that overcome resistant plants. Numerous rice varieties with different genes resistance to BPH have been produced by IRRI.

Insect-resistant transgenic plants brought about by genetic transformation currently available is generally vertical resistance with the major genes, such as Bt-gene. Anticipating the development of new biotype overcoming Bt-plants, others resistant plants with different genes should be developed.Pyramiding of major genes is a possible strategy to develop the longer useful life of resistance plants. This approach needs a reliable method to determine the presence of the two genes. If bioassay method will be adopted, several insect biotypes should be available for this purpose. The use of molecular markers is a very reliable method to diagnose the integration of the two genes in the plants. Pyramiding of major genes may be achieved through genetic transformation, for example, pyramiding the two Bt genes or combining genes encoding a toxin and a gene encoding a repellent.

Tissue-specific Expression

A great selection pressure that leads to the development of new insect biotypes cause the constitutive expression of resistance genes at all times and in the whole tissue. This problem can be dealt with by expressing the resistance traits in a specific tissue, or at certain growth stages, or only in response to insect feeding.

Disease Resistance

With transfer of genes from one organism to another becoming common place, it is possible to introduce genes conferring disease resistance into crop plants. Such gene transfers can be done by direct methods and or vector-mediated methods. Gene gun or biolistic method and *Agrobacterium*-mediated method are the best examples of the direct method and vector-mediated method, respectively.

Transgenic Plant Disease Management

Just as it was possible with coat protein-mediated plant viral resistance and with toxin-inactivating protein-mediated bacterial resistance disease resistance genes can be sourced from plant pathogens themselves. Host plants also contribute an enormous number of disease resistance genes, such as those encoding pathogenesis-related (PR) proteins, which have been used against fungal diseases.

Candidate Genes against Viral Pathogens

The best example of the use of transgenic resistance against plant diseases is one that was accomplished in the management of papaya ring spot virus (PRSV) in Hawaii. Traditional breeding in bringing about resistance against this disease posed crossability barriers. Hence coat-protein-mediated resistance using coat protein gene sourced from a Hawaiian strain of PRSV was attempted. One transgenic line was found to be completely resistant to PRSV. Transgenic resistance against banana bunchytop resistance using a BBTV replicase gene is under study. Once it becomes routine, it will be easier to deliver human vaccines (cholera toxin vaccine and hepatitis B surface antigen) via transgenic banana fruits which form an excellent delivery material.

Recently, a gene silencing mechanism has been productive in containing rice yellow mottle virus. Similar attempts also have been made in containing multiple viral infections (tomato spotted wilt virus and turnip mosaic virus) in plants.

Candidate Genes against Bacterial Pathogens

Xa21, a wide-spectrum bacterial blight resistance gene sourced from an African rice, *Oryza longistaminata,* was backcrossed into a cultivated variety by scientists of the International Rice Research Institute, the Philippines (IRRI). The resistance gene was cloned using molecular means distributed to labs worldwide so that the gene could be put into rice cultivars of local importance.

Wildfire disease of tobacco caused by *Psuedomonas syringae* pv. *tabaci* is a serious disease wherein, a phytotoxin secreted by the pathogen, drastically modifies the amino acid metabolism of the plant with the eventual accumulation of ammonia in tobacco leaves, that causes extensive blighting. Interestingly, the pathogen that synthesises the phytotoxin remains unaffected by the toxin.

Researchers carried out for a search of the candidate gene from the pathogen itself led to toxin-inactivating gene, which was named '*ttr*'. This was successfully isolated from the pathogen and the same was cloned into tobacco cultivars, which showed excellent wildfire resistance.

Candidate Genes for Fungal Resistance

PR protein genes that are a possible potential source for candidate genes for fungal resistance may play a direct role in defence by attacking and degrading the pathogen cell wall components. Typical candidate genes are that encoding chitinases and ß-1,3 glucanases. Increasing expression of individual and multiple PR-proteins in various crops have shown some success in enhancing disease resistance in particular pathogens. A chitinase gene from an anti-fungal biocontrol fungus species (*Trichoderma viride*) confers transgenic resistance against the rice sheath blight pathogen. A rice

PR-5 protein gene in wheat delays the onset of symptoms caused by the wheat scab pathogen.

Biocontrol of Plant Pathogens

The worldwide environmental concern over pesticide use has initiated a large upsurge of biological disease control. Among the various antagonists used for the management of plant diseases, *Trichoderma* and *Pseudomonas* are most useful.

Trichoderma

Among the various isolates of *Trichoderma, T. viride, T. harzianum, T.virens* and *T. hamatum* are used to control various diseases of crop plants especially with the dreaded soil-borne pathogens. As a bio-control agent it has high rhizosphere competence, ability to synthesize polysaccharide–degrading enzymes, amenability for mass multiplication, broad spectrum of action against various pathogens and, above all, its environmental friendliness.

Fluorescent Pseudomonads in Induced Systemic Resistance (ISR) against Plant Diseases

Fluorescent pseudomonads inhibit the pathogens either directly through the production of various secondary metabolites or indirectly by inducing plant-mediated defence reactions. They have the ability to colonise the rhizosphere and the persistence throughout the growing season. They are root colonisers as they occur in the natural habitat of rhizosphere and when they are reintroduced to roots through seed or seed-piece inoculation, they colonise root surface profusely. They suppress the pathogens by antibiosis through the production of various antibiotic substances such as 2,4-diacetyl phloroglucinol, phenazine-1-carboxylic acid, oomycin A, oxychlororaphine, pyoluteorin, pyrrolnitrin and pyocyanine.

Siderophores that are extracellular, low molecular weight substances which selectively complex iron with high affinity are produced by fluorescent pseudomonads.

These siderophores, such as pseudobactin and pyoverdine, chelate the iron available in the soil and make it unavailable to pathogen which due to deficiency of iron. In rice, seed treatment followed by root dipping and foliar spray with *Pseudomonas fluorescens* showed a higher induction of ISR against sheath blight pathogen, *R. solani*

Molecular Diagnostics of Plant Diseases

The foremost step in managing a plant disease is to correctly identify it. Some diseases can be diagnosed quickly by visual examination, and others require laboratory testing for diagnosis which may take days or even weeks to complete and are, in some cases, relatively insensitive. To prevent delays that can be frustrating appropriate disease control measures may be taken to prevent plant injury, especially when high value cash crops are involved. Fortunately, with advances in biotechnology, new products and techniques are becoming available that will complement or replace time-consuming laboratory procedures.

ELISA Diagnostic Kits

A number of disease detection kits, developed for use at the site where a disease is suspected, not requiring laboratory equipment, are especially useful to growers. Some tests only take five minutes to perform. The diagnostic kits uses a technique by which proteins called antibodies are used to detect disease causing organisms of plants (plant pathogens). The technique, called ELISA (enzyme-linked immunosorbent assay), is based on the ability of an antibody to recognise and bind to a specific antigen, a substance associated with a plant pathogen. The antibodies used in the diagnostic kits are highly purified proteins produced by injecting a warm-blooded animal (like a rabbit) with an antigen associated with one particular plant disease. The animal reacts to the antigen and produces antibodies that recognise and react only with the proteins associated with the causal agent of that plant disease. Colour changes on the unit's surface indicate a positive reaction. ELISA kits can detect bacterial canker of tomato and soyabean root rot.

PCR

A new technology, PCR (polymerase chain reaction), has great possibilities for enhancing the sensitivity of various assays that use nucleic acid probes. This technology is used to produce enormous numbers of copies of a specified nucleic acid sequence allows the detection of very small amounts of a pathogen in a sample by amplifying the pathogen sequences to a detectable level. PCR is especially useful in plant quarantine point of view owing to its fastness.

Virus Resistance

Many reports about the development of transgenic plants have become resistant to virus infections when they are transformed with sequences related to several gene functions.

Successes have been seen in areas related to sequences concerning viral capsid protein, viral movement protein, antisense RNA, antibody-mediated resistance, interferon-related genes and host genes involved in plant protection.

Biochemical complexes comprising an RNA or a DNA genome encased in a protein capsid which may or may not be surrounded by a membrane envelope are called viruses. The protein-coat-'covered genome is known as the nucleocapsid. The proteins on the surface of the capsid and envelope determine the interaction of the virus with the host and draw out the protective immune response against the virus. Some virus particles also contain enzymes required to facilitate the replication of the virus.

The tobacco mosaic virus (TMV) is an example of a virus with helical symmetry. Its capsomeres appear as projections that are assembled on the RNA genome into rods extending to the length of the genome. In contrast the capsomere arrange-ment in other viruses is cubical or icosahedral enclosing its nucleic acid component. Engineering resistance in plants involves either attacking the capsid properties or disrupting the virus replicating mechanisms.

Genetic Engineering for Abiotic Stress Tolerance

Abiotic stresses refer to a number of abnormal environment parameters, such as drought, salinity, cold, freezing, high temperature, anoxia, high light intensity and nutrient imbalances etc. These stresses lead to dehydration or osmotic stress through scarcity of water for vital cellular functions and maintenance of turgour pressure. Stomata closure, reduced supply of CO_2 and slower rate of biochemical reactions during prolonged periods of dehydration, high light intensity, high and low temperatures lead to high production of Reactive Oxygen Intermediates (ROI) in the chloroplasts causing irreversible cellular damage and photo inhibition.

As a result of dehydration or osmotic stress, a series of compatible osmolytes get collected for osmotic adjustment, water retention and free radical scavenging. Likewise overexpression of certain enzymes, such as superoxide dismutase, ascorbate peroxidase and glutathione reductase, is involved in free radical detoxification and scavenging of free radicals under oxidative stress.

Table 3. Genetic engineering of plants for tolerance to abiotic stresses

Stress	*Gene/Enzyme*	*Source*
Osmotic	Delta-pyrroling-5-carboxylate synthetase (P5CS)	Mothbean (*V. aconitifolia*
Drought & Salinity	Mannitol-1-phosphate dehydrogenase (*mt1D*)	*E. coli*
Cold and Salt	Choline oxidase (*cod* A)	*Arthrobactor globiformis*
Salt	Choline dehydrogenase (*bet* A)	*E. coli*
Cold	Omega-3-fatty acid desaturase (*fad* 7)	*Arabidopsis*
Drought	Trehalose-6-phosphate synthase	Yeast
Drought	Levan sucrase (Sac B)	*Bacillus subtilis*

Complexed stresses by osmoticum, dehydration and salinity

A potent and compatible osmoprotectant is proline which accumulates in high concentrations in glycophytes and halophytes in response to osmotic stress, such as drought and high salinity. Two important enzymes for the biosynthesis of proline i.e. Δ^1 - pyrroline-5-carboxylate synthetase (P5CS) and Δ^1 - pyrroline-5-carboxylate reductase (P5CR), have been cloned from several plants and their expression studied under various abiotic stresses and ABA application.Certain transgenic plants exhibiting a high level of P5CS mRNA also accumulate high level of P5CS protein, producing 10 to 18-fold more proline than the control plants. Under drought stress the proline content increased from about 80 mg/g fresh leaf (before stress) to about 3000 mg/g (after stress) in control and from 1000 mg/g to an average of 6500 mg/g in transgenic lines. Compared with the wild type (WT) control plants wilting in the transgenic plants was less severe and delayed by 2-3 days transgenic plants as compared. This shows that proline acts as an osmotic protectant and its increases production in the transgenic plants increased tolerance to drought and salt stress. The compatible solute, glycinebetaine, a quaternary amine is widely distributed among plants and protects them on exposure to salt and cold stress. In plants, like spinach and barley, betaine is synthesized from choline by catalysed by choline mono-oxygenase oxidation of choline to betaine aldehyde. In the second step it is synthesized by nuclear coded gene for betaine aldehyde dehydrogenase. The transformed plants grew slowly at 200 mM NaCl whereas none of the WT plants grew. The collection of glycine betaine through genetic engineering in Arabidopsis enhanced its ability to tolerate salt and cold stress.

Anaerobiosis/Anoxia

Most plants are highly sensitive to anoxia during submergence. Avoidance of self are all metabolic changes that are an important aspect of the adaptation to oxygen limitation poisoning and cytoplasmic acidosis, maintenance of adequate supplies of energy and sugar. During anoxia, ATP and NAD^+ are generated via glycolysis and fermentation and not in the Krebs cycle and the respiratory chain via. A number of enzymes of the anaerobic pathways such as alcohol dehydrogenase and pyruvate decarboxylase induced during anoxia have been cloned and characterised.

Heavy Metal

Since soils have elevated levels of one or more inorganic ions such as sodium in saline soils; Al, and Mn in acidic soils and heavy metals Cu, Zn Pb, Ni, Cd etc. due to mining, industrial effluents and other human activities, optimum growth and productivity and even cultivation of most of the plants is severely restricted.

Table 4. Genetic engineering of plants for tolerance to heavy metal

Stress	*Gene/Enzyme*	*Source*
Cadmium	Metallothionein-I (MT-I)	Mouse
Copper	Metallothionein-like (PsMTA)	Pea
Aluminium	Citrate synthase (CSb)	P. *aeruginosa*

Some plants are genetically resistant to Al toxicity. In such cases the Al exclusion and uptake from root tips are found to be correlated to their increased capacity to release organic acids such as citric acid which chelates Al_3^+ outside the plasma membrane. Transgenic tobacco and papaya were found to overexpress a citrate synthase gene (CSb) from *Pseudomonas aeruginosa* in their cytoplasm. Tobacco lines expressing CSb had up to 10-fold higher level of citrate in their root tissues and one of the lines released 4-fold citrate extracellularly while in papaya there was only 2 to 3-fold increase of citric acid production which was shown to result in Al tolerance in both the species.

Heat and Cold

Temperate and subtropical plants are exposed to high temperature during early tillering, flower initiation, anthesis and grain filling stages leading to which results in substantial reduction in their productivity. In response to high temperature all organisms, including plants, blend a set of proteins called heat shock proteins (HSPs) which have been classified into several families based on their molecular masses. The induction of HSPs at permissive temperatures are linked with the acquisition of thermotolerance to tolerate short periods of an otherwise lethal temperature.

The chilling sensitivity of plants is closely associated with the degree of unsaturation of fatty acids in the phosphatidylglycerol of chloroplast membranes. Plants such as spinach and Arabidopsis having a high proportion of cis-unsaturated fatty acids, are resistant to chilling, whereas species like squash with only a small proportion are not. The chloroplast enzyme glycerol-3-phosphate acyltransferase is vital for assessing the level of phosphatidylglycerol fatty acid unsaturation. This shows that the level of fatty acid unsaturation of phosphatidylglycerol and the degree of chilling sensitivity of tobacco can be controlled by transformation with cDNAs for glycerol-3-phosphate acyltransferases from squash and Arabidopsis.

Shading

While optimum supply of nutrients and efficient photosynthesis contribute to biomass production, the allocation of assimilates within the developing plant determines the harvest index and economic yield. In pure stand canopy and mixed cropping, competition for light energy enforces shade avoidance syndrome embodied by rapid growth and extension of stem and petiole at the expense of leaves, storage and reproductive organs. This leads to predisposing plants to lodging, susceptibility to diseases and insect pests and a lower harvest index.

With the advent of semi-dwarf varieties of wheat and rice in the 1960s there was higher harvest index and grain yield by countering some of the defects of the tall genotypes. However, competition among plants for light energy continues to operate in canopies under intensive cultivation practices. The photosynthetic pigments in plants absorb the visible radiation (400-700 nm) and reflect and transmit far red (FR) radiation beyond 700 nm. The FR wave band between 700 and 800 nm salient in the dense plant stands have been involved in proximity perception for initiating shade avoidance syndrome.

Photoreceptors called phytochromes, possessing distinct photo sensory functions perceive the FR reflection signals. Phytochrome (phyA) mediating the inhibition of stem growth on etiolated plants in response to FR wave length 710-720 nm is rapidly degraded and down regulated in light grown plants. Transgenic tobacco lines expressing a high level of heterologous oat phyA apoprotein have been produced.

There is correlation between the level of growth inhibition of transgenic plants and the level of phyA production. Under field trials at various planting densities, from 20 to 100 cm, the transgenic plants were identical to the WT plants at the lowest plant density but became progressively shorter as the plant density increased. This phenomenon, termed as "proximity conditional dwarfing", led to a 15 to 20 per cent increase in harvest index in transgenic plants under high plant density

This shows the suppression of shade avoidance response under high level of phyA expression. Further studies on the molecular basis of interaction of various phytochromes among themselves and with R: FR rations in natural light environment may help to change crop plant structure to avoid shade stress and obtain maximum production under high plant density, mixed cropping and agroforestry.

UV-B

The high influxes and absorption of UV-B radiation affects terrestrial plants, damaging DNA directly or indirectly through formation of free radicals, membranes by peroxidation of unsaturated fatty acids, photosystemII, phytohormones and even symbiotic relationship of plants with micro-organisms. Many secondary metabolites such as flavonoids, tannins and lignins are increased at elevated levels of UV-B radiation which screen UV-B and protect the cellular components against the UV-B damage.

Oxidative Stress

Overproduction of reactive oxygen intermediates (ROI) including H_2O_2 results in a number of abiotic stresses such as extreme temperatures, high light intensity, osmotic stresses, heavy metals and a number of herbicides and toxins causing extensive cellular damage and inhibition of photosynthesis.

Strategies for Improving Tolerance

Within a decade of the molecular understanding of pathways induced in response to one or more of the abiotic stresses the work on genetic engineering of tolerance to abiotic stresses began. In a majority of the cases the transgenes expressed faithfully but compared to the non-transformed wild type plants only a limited level of tolerance was provided under stress conditions. In many cases the transgenic plants had morphological abnormalities and slower growth under nonstressed environment. The very low level of many compatible osmolytes responsible for osmotic adjustment was not effective in providing the required water retention and osmotic adjustment.

High levels of tolerance for commercial exploitation can be achieved by the use of multiple tolerance mechanisms for one or more of the abiotic stresses through stepwise or co-transformation. The QTL mapping of stress tolerance in certain species, comparative mapping and map-based cloning in plants may be employed to screen genes which function under stress as well as those induced and expressed in response to stress.

Molecular understanding of the stress perception, signal transduction and transcriptional regulation of abiotic stress responsive genes may help in engineering tolerance for multiple stresses.

Studies on the molecular mechanism for providing protection against biotic and abiotic stresses may lead to a generalised master mechanism for stress tolerance. Optimum homeostasis is always a requisite to living organisms for adjusted environments. Thus, abiotic stress accompanying a number of biological phenomena must be precisely investigated by consideration of plant homeostasis.

Molecular Alphabets of Plant Abiotic Stress Responses

All plants have a latent ability to adjust to circadian and seasonal environmental variables which are often decisive factors in controlling certain physiological attributes. Apart from those variables, there may be certain other rapid and unpredicted disturbances in the environment resulting in stressful conditions.

For instance, water shortage for long periods due to lack of irrigation, infrequent rains or lowering of water table causes drought stress; excess water through rain, cyclones or frequent irrigation results in flooding or submergence or anaerobic stress; cultivating plants on saline soils or frequent irrigation with ground water leads to salinity stress; and sudden atmospheric heating or cooling due to transient changes in wind patterns, cloud formation or excessive sunlight causes temperature stress. Most crop plants, not selected for meeting exigencies caused by such abiotic stress factors, are unable to adjust to such conditions.

Abiotic stress affects survival, biomass production and accumulation, and grain yield of most crops. Different crop ecosystems are affected by different abiotic stress factors, and to a differential extent, with varying degrees of susceptibility. There is also some level of variation associated with specific developmental stages of the plant. Today, how plants face stress in terms of molecular alterations is the key issue in plant stress molecular biology studies.

The best studied environmental factor in plant research with respect to molecular details is both the quality and quantity of light affecting photosynthesis and growth of plants. Light is perceived through several different photoreceptors. A battery of molecular components is involved in transduction of light signal to the nuclei where it regulates transcription of selective genes, in both positive as well as negative manner. Also, light-responsive elements (LREs) present in the promoters of the lightregulated genes and the transcription factors associated with light-regulated promoters interact to produce regulated gene expression. However, as against light, there is scarcity of information on how changes in temperature, water and salt levels are perceived and translated into cellular events.

The period of 1980s onwards has been the 'Phase of Recombinant DNA Technology' with Genetic engineering as the most revolutionary tool to impact agricultural research in recent years. In the last few decades techniques of protein analysis, identification, isolation, cloning and characterisation of genes, promoter analysis, genetic transformation and new research on genomics and proteomics have contributed significantly in to plant molecular biology. This progress has led to production of a range of transgenics for varied traits. The study of genetically engineered abiotic stress-tolerant crops for high level stress tolerance has gained a great deal of attention too. There has been a lot of information on stress proteins that are specifically induced in response to abiotic stresses. Gene libraries enriched for stress-related cDNA clones have been constructed for several plant species. Availability of the genomic clones of stress genes has helped to identify the stress-related promoter sequences. Due to a higher level of sophistication achieved in the isolation of cDNA clones that are present in minute amounts in the gene libraries, identification of genes encoding transcription factors and for proteins which mediate stress signalling has become possible in selected instances. Both plant molecular biologists and plant biotechnologists feel the need to further unravel the fundamentals of the plant stress responses.

Macro- and Micro-level Stress Effects

It is generally believed that the response of plants to stress agents is particularly of adaptive nature when the stress is sublethal. In contrast, response shown may be biased towards senescence or cell death if the stress is lethal. Application of sublethal stress regimes in experimentation has become an effective approach for unveiling the fundamentals of stress responses.

Using tools of physiological, biochemical and molecular relevance, there are intensive efforts to unveil both the constitutive as well as inducible mechanisms connected with the survival of plant cells under stress conditions. While some of the stress effects are common amongst different abiotic stresses, others are unique to a particular stress type.

At the whole plant level, common effects caused by the stress factors mentioned above include reduced seed germination and seedling establishment, poor seedling vigour, decrease in the root length, leaf rolling, reduced pollen viability, leaf senescence, incomplete grain filling and reduction in grain yield. At the cellular and sub-cellular level, stress-induced unique changes include increased unsaturation of the membrane lipids in response to low temperature stress, increased levels of different osmolytes in response to osmotic factors (such as dehydration, salinity and low temperature stresses), general repression of protein biosynthesis in response to water and high temperature stresses, selective changes in K^+/Na^+ levels in response to salt stress and finally, and upregulation of glycolytic and enzymes required for alcoholic fermentation by anaerobic stress.

These cellular changes have been dealt with mostly specific probes, with a parallel development in stress studies and is the application of shot-gun approach which aims at analysing what happens in the system under sublethal stress regimes. This approach relies on more sophisticated molecular tools and techniques, such as examining stress-related changes in protein profiles and differential screening of gene libraries. This development has led to an unprecedented set of discoveries of stress proteins and stress genes which are described below.

Stress Proteins and Stress Genes

One minute of increased temperature bringing about an altered puffing pattern of the polytene chromosome in *Drosophila* has led to more research on the molecular basis of the stress responses. This showed that the heat shock (HS) conditions result in an altered protein profile in *Drosophila* cells. Soon thereafter, it was reported that HS induces comparable alterations in the protein profiles of plants too. Subsequently heat shock proteins (HSPs) have been detected and characterised in a number of plant species.

Heat shock protein (hsp) genes are the most extensively characterised of the stress genes so far, particularly in *Drosophila, Saccharomyces cerevisiae* and *E. coli.* HSPs are mostly encoded by nuclear genes, but these proteins are localised in different cell compartments, including cytoplasm, mitochondria, chloroplast and endoplasmic reticulum. Depending upon the mode of expression, broadly, hsp genes fall into the following two categories: (i) those that are constitutively expressed, often referred to as heat shock cognates (hsc), and (ii) those that are strongly induced under heat stress (hsp).

Several plant hsp genes have been cloned and sequenced since the beginning of 1980s, the most thoroughly characterised plant hsp genes being those encoding low molecular weight HSPs and HSP70. The nucleotide sequence and the structural features of the hsp are remarkably conserved. Apart from hsp genes, a large number of genes induced in response to low temperature, water, salt and anaerobic stress have been cloned and characterised in recent years.

It has been a difficult task to identify the precise physiological roles of most stress genes/proteins. The best relevant details are perhaps available for 'Anaerobic Proteins' as most of these characterise enzymes of the glycolytic and alcohol fermentation pathways. This analysis shows that respiratory pathway is affected in a major way in response to anaerobic stress. Turning to high temperature stress, a great success was achieved when biochemical and genetic tools enabled two groups to suggest that HSP70 is involved in the role of chaperoning. Thereafter, HSP60 and HSP90 have also been found important for chaperoning activities. HSP100 is shown to be critically required for resolubilising protein aggregates formed due to action of stress.

Regarding other stress responses, 'Water Stress Proteins' (WSPs) have been used in several important metabolic functions. The 'Late

Embryogenesis Abundant' or LEA proteins and osmotin may protect macromolecules and membranes, and chaperones and proteases are involved in protein turnover and protein translocation. The detoxification enzymes, such as glutathione S-transferase, catalases, superoxide dismutase and ascorbate peroxidases, are useful in protection from reactive singlet oxygen species and finally proteins involved in regulatory functions and in signal transduction, including various protein kinases and transcriptional factors have a broader role in managing stress responses. With precise information on the biochemical role(s) of stress genes/proteins being made available it should be possible to replace the operational stress gene/protein terms with more specific functional nomenclature.

Though there a lack of information on the functioning of most stress proteins/genes, work employing stress proteins of known functions as well as on those with relatively unknown functions has largely contributed in planning strategies for improving stress tolerance through transgenic technology. This stresses the fact that there is a constant need for identification, isolation and characterisation of increasing number of stress proteins as well as for unravelling the functions of those stress proteins which have not been assigned any biochemical role thus far.

Stress-Associated Genes and Proteins

Signalling Cascades and Transcriptional Control

There have been extensive studies on genes involved in signalling cascades and in transcriptional control, such as mitogen-activated protein (MAP) and salt overly sensitive (SOS) kinases, phospholipases and transcription factors. However, most of these studies deal with short-term experiments, which cannot provide conclusions about the actual stress tolerance of crops. This depends on longer term plant performance with respect to biomass, yield data and the degree of recovery from stress. It has been recently proved that constitutive expression of the tobacco mitogen-activated protein kinase kinase kinase/ Nicotinia protein kinase 1 in maize activates an oxidative signal cascade and leads to cold, heat, and salinity tolerance in the transgenic plants. The transgenic maize maintained significantly higher photosynthetic rates, indicating that NPK1 induces a mechanism that protects the photosynthetic machinery. Most current genomic research is done at the DNA and RNA levels, assuming that gene up- or down-regulation will explain abiotic stress processes. However, there has been an increasing number of demonstrations of translational and post-translational regulation of stress-associated proteins in response to environmental stress. AnnAt1 is a member of a multigene family of Ca_2^+-dependent membrane-binding osmotic stress-responsive proteins. Interestingly, its gene transcript was constitutively expressed, while protein expression was upregulated in response to stress and the protein translocated from the cytosol to the membrane. The mutant plants annAt1 and annAt4 were seen to be hypersensitive to salt and abscisic acid (ABA) during seed germination and early seedling growth.

Heat-shock Proteins and Chaperones

Plant abiotic stress tolerance involves heat-shock proteins (Hsps) and molecular chaperones, as well as late embryogenesis abundant (LEA) protein families. High temperature, salinity and drought stress can cause denaturation and dysfunction of many proteins. We thus expect that Hsps and LEA proteins help to safeguard against stress by controlling the proper folding and conformation of both structural and functional proteins. The increase in Hsp expression under conditions of abiotic stress was researched extensively by functional genomics and proteomics in different plant species. A significant osmoprotective effect was got in *Escherichia coli* transformed with the cytosolic chaperonin CCP-1a from Bruguiera sexangula. Studies on plant genetic transformation with Hsp genes have focused mostly on heat stress and thermotolerance. Overexpression of HSP101 from Arabi-dopsis in rice plants culminated in a significant improvement of growth performance during recovery from heat stress. Overexpression of LEA proteins was associated in several cases with desiccation tolerance, although the actual function of these proteins is still unknown. Recently, overexpression of HVA1, a group 3 LEA protein isolated from barley (*Hordeum vulgare* L.), conferring dehydration tolerance to transgenic rice plants was reported.

Stress-associated Changes in Metabolites and Metabolomics

Severe osmotic stress brings about adverse changes in cellular components. A wide range of metabolites that can prevent these harmful changes have been identified, including amino acids (e.g. proline), quaternary and other amines, and a variety of sugars and sugar alcohols (e.g. mannitol and trehalose). Two general plans for the metabolic engineering of abiotic stress tolerance have been put forth: increased production of specific desired compounds or reduction in the levels of needless (toxic) compounds.

However, modulation of a single enzymatic step is usually monitored by the tendency of cell systems to restore homeostasis, thus limiting the potential of this approach. Targeting multiple steps in the same pathway could help to manages metabolic fluxes in a more predictable manner.

Amino Acids

Accumulation of proline was associated with better plant performance under salt stress. Proline-level increments can be achieved in plants by overexpressing Δ^1-pyrroline-5-carboxylate synthetase (P5CS), as found, for example, in tobacco. This approach also brought about the upregulation of proline dehydrogenase, which reduces proline levels. Indeed, Arabidopsis transformation with proline dehydrogenase antisense or a knockout of this enzyme produced increased free proline accumulation and better growth performance under salt stress. To check whether plant growth reduction in response to osmotic stress might actually result from osmolyte accumulation, proline levels were manipulated by expressing mutated derivatives of P5CS from tomato in *Saccharomyces cerevisiae*. The levels of proline accumulation and cell growth were inversely correlated in cells grown under normal osmotic conditions. Alternatively, proline might confer a protective effect by generating stress-protective proteins. Exogenously applied proline and/or salt stress in *Pancratium maritimum* were found to induce the expression of ubiquitin, antioxidative enzymes, and dehydrins.

Amines

The accumulation of glycine-betaine a widely studied osmoprotectant, has been researched with respect to modifications of several metabolic steps. Betaine aldehyde decarboxylase from the halophyte *Suaeda liaotungensis* was integrated into tobacco plants and the *in vitro* plantlets were significantly resistant to salt conditions. Similar outcomes were obtained by transforming rice with the choline dehydrogenase gene (codA) from *Arthrobacter globiformis*; the gene product of codA catalyses the oxidation of choline to glycine betaine via betaine aldehyde as intermediate transgenic rice plants recovered from salt stress and set seeds, in contrast to wild-type plants.

Tomato plants transformed with a bacterial codA gene targeted to the chloroplast were highly tolerant to chilling and oxidative stress, showing an enhancement in photosynthetic rate, plant survival, flower retention and fruit set. Co-targeting multiple steps in the same pathway was a successful strategy for overexpressing glycine-betaines in plants and bacteria. Indeed, stress tolerance was increased by the genetic engineering of *E. coli* and tobacco plants with the betaine aldehyde–choline dehydrogenase fusion protein. Plant polyamines have earlier been shown to be involved in plant response to salinity. More recently, genetic engineering for increased biosynthesis of several specific polyamines resulted, in several cases, in stress-tolerant plants.

Overexpression of arginine decarboxylase (ADC), ornithine decarboxylase and S-adenosylmethionine decarboxylase produced a significant increase in putrescine levels and a small increase in spermidine and spermine levels. Transgenic rice plants expressing *Datura stramonium* ADC under the control of the monocot Ubi-1 promoter mugenerated ch higher levels of putrescine under drought stress only, promoting spermidine and spermine synthesis and ultimately protecting the plants from drought. Overexpression of spermidine synthase cDNA from *Cucurbita ficifolia* in *Arabidopsis thaliana* significantly increased spermidine levels, and consequently enhanced tolerance to various stresses.

Drought Resistance

Drought, a meteorological event, results from absence of rainfall for a period of time to cause moisture-depletion in soil and waterdeficit with a decrease of water potential in plant tissues. But from the agricultural aspect, its working definition would be lack of water availability, including precipitation and soil-moisture storage capacity, in quantity and distribution during the lifecycle of a crop plant. This restricts the expression of full genetic potential of the plant which acts as a serious limiting factor in agricultural production by preventing a crop from reaching the genetically determined theoretical maximum yield. The effect of drought on crop production and overall economy is evident in the most affected states in India which are Rajasthan, parts of Gujarat, Haryana and Andhra Pradesh. Most of the crops are sensitive to water deficits, particularly during flowering to seed development stage. Even crops grown in arid and semi-arid regions, such as pearl millet, sorghum and pigeon pea, are also affected by drought at the reproductive stage.

In agriculture, drought resistance is associated with the ability of a crop plant to produce its yield with minimum loss in a water-deficit environment relative to the water-constraint-free management. An understanding of genetic basis of drought resistance in crop plants is a prerequisite for a geneticist to sudevelop a perior genotype through either conventional breeding methodology or biotechnological approach.

Mechanisms of Drought Resistance

Genetically the mechanisms of drought resistance can be grouped into drought escape, drought avoidance and drought tolerance. However, crop plants use more than one mechanism at a time to resist drought. Drought escape is referred to as the ability of a plant to complete its lifecycle before serious soil and plant water deficits set in. This mechanism involves rapid phenological development, developmental plasticity and remobilisation of preanthesis assimilates to grain.

Drought avoidance is a plant's ability to maintain relatively high tissue water potential regardless of a shortage of soil-moisture, whereas drought tolerance is the ability to tolerate water-deficit with low tissue water potential. Improving water uptake, storing in plant cell and reducing water loss prevent drought situations. The responses of plants to tissue water-deficit determine their level of drought tolerance. Drought can be avoided by maintenance of turgour through increased rooting depth, efficient root system and increased hydraulic conductance and by reduction of water loss through reduced epidermal (stomatal and lenticular) conductance, reduced absorption of radiation by leaf rolling or folding and reduced evaporation surface (leaf area). Plants under drought condition survive by balancing between maintenance of turgour and reduction of water loss. The mechanisms of drought tolerance are maintenance of turgour through osmotic adjustment (a process which induces solute accumulation in cell), increase in elasticity in cell and decrease in cell size, and desiccation tolerance by protoplasmic resistance.

Unfortunately, adapting to drought has its own disadvantages. A genotype of short duration usually yields less compared to that of normal duration. The consequence of mechanisms that confer drought resistance by reducing water loss (such as stomatal closure and reduced leaf area) is reduced assimilation of carbon dioxide. Osmotic adjustment enhances drought resistance by maintaining plant turgour, but the increased solute concentration responsible for osmotic adjustment may have harmful effect in addition to energy requirement for osmotic adjustment. Consequently, crop adaptation must show a balance among escape, avoidance and tolerance while maintaining adequate productivity.

Genetics

Drought resistance being complex its expression depends on the action and interaction of different morphological (earliness, reduced leaf area, leaf rolling, wax content, efficient rooting system, awn, stability in yield and reduced tillering), physiological (reduced transpiration, high water-use efficiency, stomatal closure and osmotic adjustment) and biochemical (accumulation of proline, polyamine, trehalose, etc., increased nitrate reductase activity and increased storage of carbohydrate) characters. Not much is known about the genetic mechanisms that quantify these characters. The identification of genes responsible for morphological and physiological characteristics and their location on chromosome have proved futile, but their inheritance pattern and nature of gene action have been recorded. Polygenic inheritance of root characters has also been reported. The long root and high root numbers are controlled by dominant alleles, and thick root tip by recessive alleles. However, leaf rolling and osmotic adjustment have shown monogenic inheritance. Researchers have reported a drought resistance gene, *Drt1* in rice, which is related to genes for plant height, pigmentation, hull colour and awn, and has pleiotropic effect on the root system. Similarly, in cowpea drought resistance is said to be controlled by a single dominant gene.

Some more reports for other traits are available, but there is need for further investigation to better understand genetic control of morphological and physiological traits contributing to drought resistance. In addition to there changes, biochemical change involving induction of compatible solute biosynthesis is one way to bring about drought. Under drought, plants try to maintain water content by accumulating various solutes that are non-toxic and do not interfere with plant processes. Therefore, they called compatible solutes. Some of them are fructan, trehalose, polyols, glycine betaine, proline and polyamines. The various genes responsible for different enzymes involved in biosynthesis of these solutes have been identified and cloned from different organisms (bacteria, yeast, human and plant).

Breeding Approach

Three breeding approaches for drought resistance have been developed. The first is to breed for high yield under optimum (water-stress-free) condition. The basic philosophy of this is that, as the maximum genetic potential of yield is expected to be realised under optimum condition and a high positive correlation exists between performance in optimum and stress conditions, a genotype superior under optimum level will also yield relatively well under drought condition. However, the concept of expression of maximum genetic potential in optimum condition questioned is as genotype environment interaction may limit the high-yielding genotype to perform well under drought. Thus, the second approach, i.e. to breed under actual drought condition, has been suggested.

In the second approach the intensity of drought is highly variable from year to year and so environmental selection pressure on breeding materials changes drastically from generation to generation. This problem compounded with low heritability of yield makes the breeding programme complicated and slow.

Counter to the above two approaches would be to improve drought resistance in high-yielding genotypes through absorption of morphological and physiological mechanisms of drought resistance. However, transferring drought resistance in high-yielding genotypes is problematic as there is lack of understanding of the physiological and genetic basis of adaptation in drought condition. In contrast, improving the yield potential of an already resistant material may be a more paying proposition, provided there is genetic variation within such a material. Simultaneous selection in non-stress environment for yield and in drought condition for stability may be done to achieve the desired goal of generating drought-resistant genotype with high yield. As such, the breeding methodology to be applied for drought resistance is the same as that applied for other purposes. In general, pedigree and bulk method could be adopted for self-pollinated crops and recurrent selection for cross-pollinated crops. However, if transfer of a few traits relating to drought resistance to a high-yielding genotype is the goal, then backcross is the appropriate technique. On the other hand, biparental mating (half sib and full sib) preserves the broad genetic base as well as provides the scope to generate the desired genotype of drought resistance. The success of any breeding programme depends on the availability of the screening technique, especially for drought resistance.

Screening Techniques

Genetic improvement in drought resistance utilising the existing genetic variability can be achieved by an efficient screening technique, which should be fast and capable of assessing plant performance at the critical developmental stages and screening a large population with only a small sample of plant material. A combination of different traits of direct relevance, rather than a single trait, should be preferred as selection criteria. Researchers provided a framework for estimating how combination of traits influences plant water status and growth. This may usefully bridge physiology and breeding into the integrated programme of plant improvement.

The different screening techniques used so far are as follows:

i) Use of infrared thermometry for screening efficient water uptake.
ii) Banding herbicide metribusin at a certain depth of soil and use of iodine-131 and hydroponic culture under stress of 15 bar for screening root growth.
iii) Psychrometric procedure for evaluating osmotic adjustment.
iv) Diffusion porometry for leaf water conductance.
v) Use of the mini-rhizotron technique for root penetration, distribution and density in the field with minimum disturbance.
vi) Infrared aerial photography for dehydration postponement.
vii) Use of carbon isotope discrimination for selecting increased water-use efficiency.
viii) As loss of yield is the main concern for the crop plant from the agricultural aspect, plant breeders stress on yield performance under moisture-stress condition.

A drought index which gives a measure of drought based on loss of yield under drought-condition in comparison to moist condition has been employed for screening drought-resistant genotype. An artificially created water-stress environment is used to provide the condition in selecting superior genotype out of a large population. Visual scoring or measurement for maturity, leaf rolling, leaf length, angle, root morphology and other morphological characters of direct relevance to drought resistance are also considered.

Biotechnological Approach

The techniques for gene transformation of crop plants have been used to identify genes responsible for drought resistance and their transfer. Mainly two approaches, namely targeted and shotgun approaches, make it possible for genetic engineering to obtain transgenic plants conferring drought resistance.

Targeted Approach

Metabolic pathways that synthesize different metabolites such as polyamine, carbohydrate, proline, glycine betaine and trehalose are associated with drought resistance. This approach depends on relevant information on biochemical reaction available for synthesis of these metabolites, and uses the related genes to transfer them from different sources to crop plants. This approach being more precise and methodical, has a higher probability of success when compared to the shotgun approach.

Lately, introducing drought-induced genes involved in different biochemical pathways from different sources to sensitive plants has emerged as one of the promising methods. The gene *TPS1* found in yeast encodes for trehalose-6-phosphate synthetase and is involved in biosynthesis of trehalose. Tobacco has been transformed with the yeast *TPS1* gene. By determining the water loss of detached leaves or the effect of withholding irrigation on the death and damage of leaf, it is evident that the transgenic plants have increased drought resistance.

The gene *P5CS* encodes for pyrroline-5- carboxylate synthetase, involved in proline synthesis, and the over-production of proline provides drought resistance. Transgenic tobacco over-expressing *P5CS* gene transferred from mothbean showed a high level of enzyme and produced 10–18-fold more proline than control plant. The over-production of proline encouraged root biomass and flower development under drought condition.

The bacterial gene *SacB,* found in *Bacillus subtilis,* encodes for levan sucrase, which has a role in fructan synthesis. When this gene was transferred to tobacco, the transgenic plant produced fructan and performed better in comparison to control under PEG-mediated drought condition.

The genes *betA* encoding for choline dehydrogenase and *betB* encoding for betaine aldehyde dehydrogenase are implicated in the biosynthesis of glycine betaine, and accumulation of glycine betaine provides drought resistance. Holmstrom *et al.* transferred *betB* gene from *Escherichia coli* to tobacco.

With the availability of genes involved in polyamine biosynthesis such as *ADC* (encodes for arginine decarboxylase), *ODC* (encodes for ornithine decarboxylase) and *SAMDC* (encodes for S-adenosyl-methionine decarboxylase), it is now easier to manage polyamine content using sense and antisense constructs of these genes in transgenic plants. Transgenic tobacco plants with *ODC* gene from yeast and mouse, *ADC* gene from oat and *SAMDC* gene from human have been recorded, yet more needs to be done to observe whether the transgenics show any tolerance to drought. Only the expression level of polyamines in plants has been studied. However, *SOD* (superoxide dismutase) gene from pea has been transferred to tobacco and transgenics were found to be drought resistant.

References

Donnellan, Craig. *Genetic Modification (Issues)*. Independence Educational Publishers. 2004.

Morgan, Sally. *Superfoods: Genetic Modification of Foods (Science at the Edge)*. Heinemann. 2003.

Smiley, Sophie. *Genetic Modification: Study Guide (Exploring the Issues)*. Independence Educational Publishers. 2005.

Roelofs, D. et al. "Functional ecological genomics to demonstrate general and specific responses to abiotic stress." *Functional Ecology* 22: 8–18, 2008.

Wang, W., Vinocur, B. and Altman, A. "Plant responses to drought, salinity and extreme temperatures towards genetic engineering for stress tolerance." *Planta* 218: 1-14, 2007.

5

Transgenic Plants as Bioreactors

Plants can be used as cheap chemical factories that require only water, minerals, sun light and carbon dioxide to produce thousands of sophisticated chemical molecules with different structures. By transferring the right genes, plants can serve as bioreactors to modified or new compounds such as amino acids, proteins, vitamins, plastics, pharmaceuticals (peptides and proteins), drugs, enzymes for food industry and so on. The transgenic plants as bioreactors have some advantages such as the cost of production is low, there is an unlimited supply, safe and environmental friendly and there is no scare of spread of animal borne diseases.

Tobacco is the most preferred plant as a transgenic bioreactor because it can be easily transformed and engineered. Tobacco is an excellent biomass producer with about 40 tons of fresh leaf production as against e.g. rice with 4 tons. The seed production is very high (approx. one million seeds per plant) and it can be harvested several times in a year.

Some of the uses of transgenic plants are:

Improvement of Nutrient Quality

Transgenic crops with improved nutritional quality have already been produced by introducing genes involved in the metabolism of vitamins, minerals and amino acids.

A transgenic Arabidopsis thaliana that can produce ten-fold higher vitamin E (alpha-tocopherol) than the native plant has been developed. The biochemical machinery to produce a compound close in structure to alpha-tocopherol is present in A. thaliana. A gene that can finally produce alpha-tocopherol is also present, but is not expressed. This dormant gene was activated by inserting a regulatory gene from a bacterium which resulted in an efficient production of vitamin E.

Glycinin is a lysine-rich protein of soybean and the gene encoding glycinin has been introduced into rice and successfully expressed. The transgenic rice plants produced glycinin with high contents of lysine.

Using genetic engineering Prof Potrykus and Dr. Peter Beyer have developed rice which is enriched in pro-vitamin A by introducing three genes involved in the biosynthetic pathway for carotenoid, the precursor for vitamin A. The aim was to help millions of people who suffer from night blindness due to Vitamin A deficiency, especially whose staple diet is rice. The presence of beta-carotene in the rice gives a characteristic yellow/orange colour, hence this pro-vitamin A enriched rice is named as Golden Rice.

The genetic engineering is also being used to improve the taste of food e.g. a protein 'monellin' isolated from an African plant (Dioscorephyllum cumminsii) is about 100,000 sweeter than sucrose on molar basis. Monellin gene has been introduced into tomato and lettuce plants to improve their taste.

Improvement of Seed Protein Quality

The nutritional quality of cereals and legumes has been improved by using biotechnological methods. Two genetic engineering approaches have been used to improve the seed protein quality. In the first case, a transgene (e.g. gene for protein containing sulphur rich amino acids) was introduced into pea plant (which is deficient in methionine and cysteine, but rich in lysine) under the control of seed-specific promoter. In the second approach, the endogenous genes are modified so as to increase the essential amino acids like lysine in the seed proteins of cereals.

These transgenic routes have helped to improve the essential amino acids contents in the seed storage proteins of a number of crop plants. E.g. overproduction of lysine by de-regulation. The four essential amino acids namely lysine, methionine, threonine, and isoleucine are produced from a non-essential amino acid aspartic acid. The formation of lysine is regulated by feed back inhibition of the enzymes aspartokinase (AK) and dihydrodipicolinate synthase (DHDPS). The lysine feedback- insensitive genes encoding the enzymes AK and DHDPS have been respectively isolated from E. Coli and Cornynebacterium. After doing appropriate genetic manipulations, these genes were introduced into soybean and canola plants. The transgenic plants so produced had high quantities of lysine.

Diagnostic and Therapeutic Proteins

Experiments are going on to use transgenic plants in diagnostics for detecting human diseases and therapeutics for curing human and animal diseases. Several metabolites and compounds are already being produced in transgenic plants e.g. the monoclonal antibodies, blood plasma proteins, peptide hormones, cytokinins etc. The use of plants for commercial production of antibodies, referred to as plantbodies, is a novel approach in biotechnology. The first successful production of a functional antibody, namely a mouse immunoglobulin IgGI in plants, was reported in 1989. This was achieved by developing two transgenic tobacco plants-one synthesizing heavy chain gamma- chain and other light kappa- chain, and crossing them to generate progeny that can produce an assembled functional antibody. In 1992, C.J. Amtzen and co-workers expressed hepatitis B surface antigen in tobacco to produce immunologically active ingredients via genetic engineering of plants.

Several other therapeutic proteins have also been produced like haemoglobin and erythropoietin in tobacco plants, lactoferrin in potato, trypsin inhivitor in maize etc. The first proteins/enzymes that were produced in transgenic plants (maize) are avidin and beta-glucuronidase and are used in diagnostic kits.

Edible Vaccines

Crop plants offer cost-effective bioreactors to express antigens which can be used as edible vaccines. The approach is to isolate genes encoding antigenic proteins from the pathogens and then expressing them in plants. Such transgenic plants or their tissues producing antigens can be eaten for vaccination/immunization (edible vaccines). The expression of such antigenic proteins in crops like banana and tomato are useful for immunization of humans since banana and tomato fruits can be eaten raw.

Transgenic plants (tomato, potato) have been developed for expressing antigens derived from animal viruses e.g. rabies virus, herpes virus. In 1990, the first report of the production of edible vaccine (a surface protein from

Streptococcus) in tobacco at 0.02% of total leaf protein level was published in the form of a patent application under the International Patent Cooperation Treaty (Mason and Arntzen,1995).The first clinical trials in humans, using a plant derived vaccine were conducted in 1997 and were met with limited success. This involved the ingestion of transgenic potatoes with a toxin of E. coli causing diarrhea.

The process of making of edible vaccines involves the incorporation of a plasmid carrying the antigen gene and an antibiotic resistance gene, into the bacterial cells e.g. Agrobacterium tumefaciens. The small pieces of potato leaves are exposed to an antibiotic which can kill the cells that lack the new genes. The surviving cells with altered genes multiply and form a callus. This callus is allowed to grow and subsequently transferred to soil to form a complete plant. In about a few weeks, the plants bear potatoes with antigen vaccines.

The bacteria E.coli, V. cholerae cause acute watery diarrhea by colonizing the small intestine and by producing toxins. Chloera toxin (CT) is very similar to E.Coli toxin. The CT has two subunits, A and B. Attempt was made to produce edible vaccine by expressing heat labile enterotoxin (CT-B) in tobacco and potato.

Another strategy adopted to produce a plant-based vaccine, is to infect the plants with recombinant virus carrying the desired antigen that is fused to viral coat protein. The infected plants are reported to produce the desired fusion protein in large amounts in a short duration. The technique involves either placing the gene downstream a subgenomic promoter, or fusing the gene with capsid protein that coats the virus.

Advantages of Edible Vaccines

The edible vaccines produced in transgenic plants will sole the storage problems, will ensure easy delivery system by feeding and will have low cost as compared to the recombinant vaccines produced by bacterial fermentation. Vaccinating people against dreadful diseases like cholera and hepatitis B, by feeding them banana, tomato, and vaccinating animals against important diseases will be an interesting development.

Biodegradable Plastics

Polythenes and plastics are one of the major environmental hazards. Efforts are on to explore the possibility of using transgenic plants for biodegradable plastics. Transgenic plants can be used as factories to produce biodegradable plastics like polyhydroxy butyrate or PHB. Genetically engineered Arabidopisis plants can produce PHB globules exclusively in their chloroplasts without effecting plant growth and development. The large-scale production of PHB can easily be achieved in plants like Populus, where PHB can be extracted from leaves.

Molecular Breeding

The term molecular breeding is frequently used to represent the breeding methods that are coupled with genetic engineering techniques. Up till now, conventional breeding methods have been used to meet the food demands of the growing world population and the challenges of poverty and improved crop production and yields. However in the years to come, the development in the agriculture yields and techniques is going to be due to the use of molecular breeding programme.

Linkage analysis which deals with the studies to correlate the link between the molecular marker and a desired trait is an important aspect of molecular breeding programme. In the past, linkage analysis was carried out by use of isoenzymes and the associated polymorphisms. Now a days, molecular markers are being used.

Molecular breeding involves breeding using molecular (nucleic acid) markers. A molecular marker is a DNA sequence in the genome which can be located and identified therefore molecular markers can be used to identify particular locations in the genome.

Due to mutations, insertions, deletions, etc. the base composition at a particular location may be different in different plants. These differences, termed polymorphisms, allow DNA markers to be mapped in a genetic linkage group.

Generally, there are three types of markers used in screening/selection:

a) Morphological marker based on visible character (phenotypic expression) e.g. flower color, seed color, height, leaf shapes, etc. Morphological markers could be dominant or recessive. There are certain constraints in using these markers as the morphological markers are easily influenced by environmental factors and thus may not represent the desired genetic variation. Some of the visible markers have not much role to play in the plant breeding programme.

b) Biochemical marker: The proteins produced by gene expression are also used as markers in plant breeding programmes. The most commonly used are isozymes, the different molecular forms of the same enzyme. Each individual variety has its own isozyme variability (profiles) which can be detected by electrophoresis on starch gel.

c) Molecular marker based on DNA polymorphism detected by DNA probes or amplified products of PCR, e.g.Restriction fragment length polymorphism (RFLP), Randomly Amplified polymorphic DNA (RAPD), variable Number Tandom Repeats (VNTR), Microsatellites, etc. Plant breeders always prefer to detect the gene as molecular marker, although it is not always possible. Molecular markers provide a true representation of the genetic make up at the DNA level. They are consistent and free from environmental factors, and can be detected much before the development of plants occur. The advantage with a molecular marker is that a plant breeder can select a suitable marker for the desired trait which can be detected well in advance. A large number of markers can be generated as per the needs. The molecular markers to be used in plant breeding programme should have the following characteristics:

   (i) the marker should be closely linked with the desired trait,

   (ii) the marker screening methods should be effective, efficient, reproducible and easy to carry out,

   (iii) the entire analysis should be cost effective.

Molecular makers are of two types:

(a) based on nucleic acid (DNA) hybridization- This involves the cloning of the DNA piece followed by the hybridization with the genomic DNA, which is later detected. The Restriction fragment length polymorphism (RFLP) was the very first technology employed for the detection of polymorphism, based on the DNA sequence differences. RFLP is mainly based on the altered restriction enzyme sites, as a result of mutations and recombinations of genomic DNA. The procedure involves the isolation of genomic DNA and it's digestion by restriction enzymes. The fragments are separated by electrophoresis and finally hybridized by incubating with cloned and labeled probes.

(b) Molecular markers based on PCR amplification. Polymerase chain reaction (PCR) is a novel technique for the amplification of selected regions of DNA. The most important advantage is that even a minute quantity of DNA can be amplified and the PCR- based molecular markers require only a small quantity of DNA to start with. Random amplified polymorphic DNA (RAPD) markers use PCR amplification where the DNA is isolated from the genome and is denatured. The template molecules are annealed with primers and amplified by PCR. The amplified products are separated on electrophoresis and identified. Based on the nucleotide alterations in the genome, the polymorphisms of amplified DNA sequences differ which can be identified as bends on gel electrophoresis.

Amplified fragment length polymorphism (AFLP) is a novel technique involving a combination of RFLP and RAPD. AFLP is based on the principle of generation of DNA fragments using restriction enzymes and oligonucleotide adaptors (or linkers), and their amplification by PCR.

Molecular Farming and Industrial Applications

With the advent of biotechnology, agricultural crops and livestock can be utilised for entirely new purposes. Just as genetically engineered bacteria are routinely being used for economical, efficient production of a wide range of medicinal proteins and industrial enzymes, plants and animals may be engineered to produce a variety of valuable biological molecules ranging from medicinals such as vaccines to polymers such as biodegradable plastics.

Molecular farming has many advantages, one of which is that biologically derived substances are most efficiently produced through biological means, utilising natural and renewable resources and ensuring optimal environmental compatibility. This will lead to the creation of new markets for farmers which complement their existing business.

The "proof of concept" exists in the form of numerous examples of transgenic plants comprising a variety of new products in their leaves or seeds, and transgenic animals producing therapeutic proteins in their milk. Some changes in production methods may be necessary, as with the espousal of any new plant or animal as an agricultural commodity; for example, tobacco producing new materials in the leaf might be machine-harvested green before extraction and purification of the product. However, it is expected that molecular farming will relate directly to traditional agricultural skills and resources. There has recently been transition from "experimental" status to profitable opportunities for farmers.

Until recently, improvement in an agricultural enterprise depended largely on breeding programmes to enhance the productive capacity of crops and livestock. But now genetic engineering has introduce entirely new characteristics very efficiently, by introducing the gene(s) coding for the desired characteristic directly into the genetic makeup of a plant or animal. This technology uses natural resources and is both specific and precise only a defined minimal set of genes is introduced into the host. Genetic engineering is thus far more efficient than traditional breeding. Additionally, the ability to introduce non-native genes allows plants and animals to produce entirely novel materials for innovative applications.

Plants, in particular, for long have natural chemicals and materials, such as flavourings, fragrances, medicines, dyes, rubber and oils. By adding or deleting genes from these plants, it is possible to alter the traits of the products or create entirely new ones. For example, rapeseed (canola) oil, normally used for cooking and margarine production, has been improved upon not only to be healthier for cooking but also for use in manufacturing detergents and novel lubricants. Soyabeans have also been modified for novel bio-based lubricants that are renewable and fully biodegradable, having fewer risks for humans and the environment in shipping, handling, and production. In some certain applications, these lubricants may be competitive with petrochemical-derived lubricants. In another example, tobacco modified to manufacture several products including antibiotics, novel polymers, a dental treatment, and anti-cancer drugs. Dozens of compounds have been produced in a whole multitude of crop plants ranging from traditional commodity crops like corn, soyabeans and tobacco to special crops like spinach, potatoes and beets. Several pharmaceuticals made in plants thus are in late-stage clinical trials and are soon to hit the market. The possibilities of plant molecular farming for these types of pharmaceutical and industrial proteins are considered huge because of the ease of boosting up their production from experimental levels to field production and their estimated low production costs. Some highly valuable pharmaceuticals that are required only in small doses will likely be manufactured on a very small scale (e.g., in a greenhouse). Other products such as biodegradable plastics, which are presently at an initial stage of development, may eventually require thousands or even millions of acres for production.

With considerable interest in livestock-based production systems, goats, pigs, cows, sheep, and chickens have all been engineered to produce a variety of new proteins, mostly for the medical industry. Advantages accruing from these systems are their relatively low operating costs and unlimited scale-up. Another benefit of the animal-based systems is that the desired protein can be produced in the milk of mammals or the eggs of thereby allowing easier extraction and purification. As an example, one recent development in animal molecular farming is the scale-up of a herd of goats engineered to produce a spider's web protein in their milk. This "biosteel" will revolutionise the materials industry, as it is both stronger and more flexible than steel, and offers a lightweight alternative to carbon fibre. Potential uses include applications for which strength and lightness are essential, such as aircraft, racing vehicles and bulletproof clothing, and for artificial tendons, ligaments and limbs. In addition, 40 to 50 pharmaceuticals are already being developed for production in animal systems; at least two reached the late-stage clinical trials in the summer of 2000.

Economic Viability of Molecular Farming Systems

The economic viability of molecular farming systems cannot be generalised as each application is different, with variation in farming practices, efficiency of the gene-expression system, ease with which the product can be extracted and purified, and prospects for obtaining useful by-products. Though the plant systems are different from animal systems, there are extreme contrasts even among the plants and among the animals. For example, tobacco is a very different crop compared to corn, soyabeans or canola, as it has the advantage of producing large quantities of green leaf material per acre, and convenient to work with from a biotechnology aspect. Hence it may be the ideal "factory" for products produced in green leaves. In cases where the products need to be produced in seeds, a better option may be corn, soyabeans, or canola because tobacco seeds are extremely small. All crops have different genetic composition, different characteristics and different production methods, making each a unique "vehicle" for molecular farming.

These differences make it impossible to establish the "cost" of generating a product using molecular farming. However, certain principles are quite clear: for both plants and animals, profitability requires high expression of the introduced gene, maintenance of product integrity, and the ability to boost up, harvest, recover, purify, and store the target product as economically as possible. For products that require high purity, the processing and purification costs are expected to be very high.

Today, in the commercial scale-up of molecular farming several suitable products have been identified, especially in the medical field, and the technology is being perfected for large-scale implementation. The focus first is on high-value pharmaceuticals molecular-farmed owing to their high profit potential. Several such products are under development and both large and small companies. For success in the competitive marketplace, cost cutting will be crucial, especially in the area of purification and product recovery. Ultimately, if the products and processes of molecular farming are to be commercially successful, they must have a competitive edge over existing, alternative products and processes (e.g., pharmaceuticals produced through microbial fermentation). Or, if the products are entirely new (e.g., biodegradable plastics or "biosteel"), there must be a parallel new market for them. With this technology continuing to develop and molecular farming production becoming more effective larger-scale, lower-value products such as industrial chemicals or biodegradable plastics should lead to larger-scale opportunities for agricultural producers.

Ensuring environmental and process regulations being developed for this industry, it will be necessary to set a regulatory regime that balances safeguarding the public and the environment and sustaining the innovative process. Though the commercial progress is advancing, molecular farming is still very much a nascent industry. Contract production is likely, and profit margins may differ considerably depending on the value of the target and the increased production requirements. But what is unique about molecular farming is that this opening should provide a captive new market for agricultural products, one that can increase over time and one that provides the pioneer producers with a clear advantage.

Host Systems and Expression Technology

In the recent past plant-based expression systems have mushroomed as a serious competitive force in the large-scale production of recombinant proteins. The first plant-derived technical proteins have already hit the market, and detailed economic estimates have shown their viability against established market sectors. Now, several plant-derived biopharmaceutical proteins are nearing the penultimate stages of commercial development. For these products that include antibodies, vaccines, human blood products, hormones and growth regulators, plants offer practical and safety advantages as well as lower production costs compared with traditional systems based on microbial or animal cells, or transgenic animals. With several products in development, molecular farming in plants is gaining importance. In this, recent technological developments in molecular farming focus on plant host systems and expression technology.

Production of Recombinant Proteins

The comparatively low cost of large-scale production a major advantage makes transgenic plants for molecular farming a great asset. Both capital and running costs are markedly lower than those of cell-based production systems since fermenters or the skilled personnel to run them are eliminated. It is estimated that recombinant proteins can be produced in plants at 2–10 per cent of the cost of microbial fermentation systems and at 0.1 per cent of the cost of mammalian cell cultures, depending on the product yield. For high-yielding proteins, the economic advantages of plant production systems are clear. One bushel of maize producing recombinant avidin at 20 per cent total soluble seed protein has the same total yield as one tonne of chicken eggs – the natural source of avidin – but at 0.5 per cent of the cost. Not many proteins can be produced at this level, but yields of 0.1–1.0 per cent total soluble protein (TSP), the typical levels necessary for the production of pharmaceutical proteins, such as recombinant antibodies, are sufficiently competitive with other expression systems to make plants economically feasible. Where such yields cannot be gained, molecular farming in plants might not be viable particularly if the product has a low market value.

For any given expression system, scalability is a consequential commercial advantage. Fermentation systems and transgenic animals have definite potential in this respect, whereas in plant-based production the scale can be modulated quickly in response to market demand simply by using more or less land as per demand. The speed of scale-up is also important as it can take several years to achieve tenfold boost in a host of transgenic sheep using natural breeding cycles, but a field of transgenic plants can be scaled-up more than 1000-fold in a single generation due to the fertile seed output. Where scale-down is required, surplus transgenic animals have to be forsaken or maintained at a loss, whereas the acreage of land planted to a specific transgenic crop can be adjusted as needed, and transgenic plant lines can be stored indefinitely and inexpensively as seed.

The costs of production of a recombinant protein also depend to a large extent on the required purity because.85 per cent of expenditure show downstream processing rather than production as such. The same methods are used for protein purification despite of the expression system, and the costs of processing are similar across the board when high purity is needed. However, plants are beneficial as several types of recombinant protein can be used in unprocessed or partially processed material, therefore eliminating many of the downstream costs. For example, recombinant subunit vaccines procured from plants can be dispensed by the consumption of raw or part-processed fruits and vegetables, and antibodies for passive immunisation can be given as topically applied pastes following minimal purification. Similarly, industrial enzymes such as glucanase and phytase can be integrated into the industrial process either as part-processed plant material or expressed directly in the plant that needs to be processed.

For purifying pharmaceutical proteins before use, several techniques have been developed to bring down downstream processing costs. As the cost of processing is inversely proportionate to the concentration of product in the starting material, the yield of protein per unit of plant biomass is imperative, depending on both the expression system and the product itself. This needs to be evaluated on an empirical basis. One of the benefits of recombinant protein expression in the seeds of transgenic cereal plants is that high levels of the product can collect in a small volume, which reduces the costs related to processing. Where conventional extraction from seeds is too costly, further strategies to aid in purification can be used. An example is the oleosin-fusion platform in which the target recombinant protein is expressed in oilseed crops as a blending with oleosin. The fusion protein can be recovered from oil bodies using a simple extraction procedure, and the recombinant protein is isolated from its fusion partner by endoprotease digestion. Similarly, a strategy is devised in which recombinant proteins are expressed as fusion constructs comprising an integral membrane-spanning domain procured from the human T-cell receptor. The recombinant protein accumulates at the plasma membrane and can be extracted in a small volume using appropriate buffers and detergents.

Exploiting the ability of vegetative tissues (e.g. leaves or roots) is another system to secrete recombinant proteins in their exudates, enabling proteins to be collected continuously. Such an approach for the production of human secreted alkaline phosphatase is used for the synthesis of recombinant antibodies, but is suitable only for small-scale production in high-containment facilities. This is because field deployment would result in large amounts of recombinant protein leaching into the soil and groundwater.

Development Timescale

The time required for the gene-to-protein process for transgenic plants includes the preparation of expression constructs, transformation, regeneration and the production and experimentation of several generations of plants. The testing phase is essential to ensure the soundness of transgene and expression, and the biochemical activity of the product, as well as the absence of adverse phenotypic changes in the host plant. These processes may take up to two years depending on the plant species, although milligram amounts of protein might be available after several months for initial testing. It is more beneficial for microbial expression systems and animal cells here as they can produce the first batches of recombinant protein more quickly than plants, but this might be outbalanced by their limitations in terms of overall scalability and cost. In addition many plant-based transient expression systems are available for the quick production and testing of recombinant proteins on a small scale, and these can be used to assess expression constructs and product quality before relegating to the expense of transgenic plants. Plant-cell-suspension cultures also produce recombinant proteins more rapidly than transgenic plants because the development and testing schedule is much shorter. Suspension cells are therefore being used for molecular farming, especially where high containment is an advantage.

Product Authenticity

In the production of pharmaceutical proteins, plants are viewed as being much safer than both microbes and animals as they generally lack human pathogens, oncogenic DNA sequences and endotoxins. However, it is important to weigh the intrinsic safety of plant-derived human proteins because the structural authenticity of such proteins affects their behaviour *in vivo*. The protein synthesis pathway is conserved between plants and animals, so plants appear to fold and assemble recombinant human proteins efficiently, a great advantage over bacterial expression systems, in which many proteins fail to fold properly and are therefore, degraded (resulting in low yields) or accumulate as insoluble inclusion bodies. The efficiency of plants to fold and assemble complex proteins correctly is proved by their capacity to produce functional serum antibodies, which contain four polypeptide chains covalently joined by disulphide bonds. More remarkably, plants produce functional secretory antibodies, which are dimers of the typical serum immunoglobulins and have two additional polypeptide components, making ten separate polypeptides in total. Two different cell types are needed to assemble such antibodies in mammals.

Although plants and animals have the same the protein synthesis pathway in they have some differences in posttranslational modifications, particularly with respect to glycan-chain structure. Thus, plant-derived recombinant human proteins tend to have the carbohydrate groups $\beta(1 \rightarrow 2)$-xylose and $\alpha(1 \rightarrow 3)$-fucose, which are absent in mammals, but do not have the terminal galactose and sialic acid residues that are found on many native human glycoproteins. Compared with other production systems it is clear that even minor differences in glycan structures can change the distribution, activity or longevity of recombinant proteins in contrast with their native counterparts, and could render such proteins immunogenic when administered to humans. Several modifications in the glycosylation pathway are needed to produce proteins with typical human glycan structures in plants. For this, the strategies to 'humanise' the glycan structures of recombinant human glycoproteins include: (i) the use of purified human β(1,4)-galactosyltransferase and sialyltransferase enzymes for the *in vitro* modification of plant-derived recombinant proteins; and (ii) expression of human b(1,4)-galactosyltransferase in transgenic plants to produce recombinant antibodies with galactose-extended glycans. In the case of the latter, 30 per cent of the antibody was galactosylated, which is similar to the proportion found in hybridoma cells. *In vivo* sialylation will be more difficult to achieve because plants lack the precursors and metabolic capability to produce this carbohydrate group. Although these methods involve the addition of carbohydrate groups missing in plants, there is also the difficulty of removing plant-specific carbohydrates. Inhibition of the enzymes fucosyltransferase and xylyltransferase using antibodies, ribozymes orRNA interference can be used to achieve this, and gene targeting by homologous recombination has produced recombinant proteins lacking plant-specific glycans in the moss *Physcomitrella patens*.

Although achieving authenticity should be the target in molecular farming, it is worth noting that studies in which mice were administered a recombinant antibody containing plant-specific glycans showed no proof of an immune reaction.

Maximising Yields in Transgenic Plants

Establishing the commercial viability of molecular farming is one of the most important factors in the achievement of adequate recombinant protein yields. Absolute yields depend on the species used for production but the selection of production crop depends on several factors in addition to yield potential.

High-level Transgene Expression

Expression-construct design can help in procuring high yields in transgenic plants by maximising the rates of transcription and translation. For dicotyledonous species, the strong and constitutive cauliflower mosaic virus 35S (CaMV 35S) promoter is often selected to get transgene expression. In cereals, where the CaMV35S promoter has a lower activity, themaizeubiquitin-1 (ubi-1) promoter is preferred. The rate of transcription in cereals is also enhanced by the addition of an intron; this phenomenon is known as intron-mediated enhancement.

Regulated promoters have both practical and biosafety advantages and hence can be used in preference to constitutive promoters. For example, although constitutive promoters enable high-level accumulation of recombinant proteins in seeds, the proteins are also expressed in leaves, pollen and roots. This might affect the growth and development of vegetative parts of the plant negatively and could expose herbivores, pollinating insects and microbes in the rhizosphere to the effects of the recombinant protein. Restriction of protein accumulation to seeds helps to offset these risks. Inducible promoters can be availed of to limit recombinant protein expression to just before or following harvest, as has been shown with recombinant glucocerebrosidase expressed in tobacco.

The rate of translation can be increased by establishing that any mRNA instability sequences are removed from the transgene construct, and that the translational startsite such the Kozak consensus for plants. Modification might be need to codon usage in some transgenes, not only to maximise the rate of protein synthesis, but also to freeze out cryptic introns and instability sequences.

There can be maximisation of both transcription and translation can be maximised by taking precautions against transgene silencing, a group of epigenetic phenomena that can restrain the expression of even structurally intact transgenes at either the transcriptional or post-transcriptional level. Strong triggers for silencing encompass prokaryotic DNA sequences and sequences that have the potential to form hairpin secondary structures at the DNA level, or double-stranded RNA when expressed. Useful techniques to minimise the frequency of silencing include the use of clean DNA procedures to escape vector backbone integration.

Protein Targeting

Targeting of recombinant proteins to oil bodies or the plasma membrane is a useful strategy to facilitate isolation and purification. However, subcellular targeting can be a general approach to increase the yield of recombinant proteins as the compartment in which a recombinant protein collect strongly affects the interrelated processes of folding, assembly and post-translational modification. All of these contribute to protein stability and, hence, help to determine the final yield.

It is more advantageous for high-level protein accumulation comparative targeting experiments with full-size immunoglobulins and single-chain Fv fragments which have shown that the secretory pathway is a more suitable compartment for folding and assembly than the cytosol. Since many plant-derived recombinant proteins under development are human proteins that normally pass through the endomembrane system, this principle can be applied not only to antibodies but also more generally. Antibodies directed to the secretory pathway using either plant or animal N-terminal signal peptides usually collect to levels that are extremely greater than those of antibodies expressed in the cytosol. Rare exceptions to this generalisation indicate that intrinsic features of each antibody might also contribute to overall stability.

The endoplasmic reticulum (ER) contributes an oxidising environment and a profusion of molecular chaperones, while there are few proteases. These are probably the most important factors influencing protein folding and assembly. Recently research, shows that antibodies targeted to the secretory pathway in transgenic plants interact specifically with the molecular chaperone BiP (binding protein).

Due to lack of further targeting information, the expressed protein is secreted to the apoplast. Depending on its size, the protein can be kept therein or might leach from the cell, with important implications for production systems based on cell-suspension cultures. The stability of antibodies in the apoplast is lower than in the lumen of the ER. Therefore, antibody expression levels can be enhanced even more if the protein is regenerated to the ER lumen using an H/KDEL C-terminal tetrapeptide tag. Accumulation levels are generally 2–10-fold greater compared with an identical protein lacking the H/KDEL signal. As an added advantage antibodies recovered in this way are not modified in the Golgi apparatus, suggesting they have high-mannose glycans but not plant-specific xylose and fucose residues. Again, although the principles of ER-retention in molecular farming have been determined using antibodies, it is likely that they will also apply to many other proteins. High levels of protein expression have also been shown when transgenes are introduced into the chloroplast rather than the nuclear genome. Several examples of chloroplast-based molecular farming in tobacco have been reported. Here the technology for chloroplast gene transfer that is most advanced include: (i) production of human growth hormone at levels approaching 8 per cent TSP; (ii) production of human serum albumin at levels exceeding 11 per cent TSP; (iii) production of cholera and tetanus toxin fragments at up to 25 per cent TSP; and (iv) production of a thermostable xylanase at 6 per cent TSP, with 85 per cent recovery.

The high turnover with the chloroplast transgenic system is advantageous, and chloroplasts also offer biosafety advantages in terms of transgene containment, due to maternal inheritance. However, the use of this system as a general approach in molecular farming is restrained by the inability of chloroplasts to carry out many post-translational modifications, including glycosylation, and biosafety concerns have been raised by the recent demonstration of horizontal gene transfer from the chloroplasts of transplastomic plants tobacteria, under laboratoryconditions. An alternative approach is to express the protein from the nuclear genome but initiate a chloroplast targeting sequence. In an interesting recent report, researchers expressed a camelid heavy-chain antibody in potatoes, and succeeded in targeting the protein to the chloroplast where it suppressed the starch-branching enzyme. The aim of the study was metabolic engineering rather than molecular farming, but that camelid heavy-chain antibodies are extremely stable makes these molecules useful candidates for both medical and industrial applications.

Pharmaceutical Applications

Pharmaceutical therapeutics are which have medical applications, can be classified into three broad categories — antibodies, vaccines and 'other therapeutics'.

Antibodies

While antibodies are proteins produced by the immune system in response to the presence of a specific antigen, an antigen is any substance, for example, a bacterial or viral protein, which is capable of causing an immune response. Monoclonal antibodies (mAbs) are specific antibodies that have been produced in a laboratory, purified and which can recognise single antigens. Certain mAbs for human tare useful herapeutic purposes. Traditionally, such mAbs were derived from mice and cell cultures, but this system has several disadvantages, including the high set-up and running costs, the limited opportunities to scale up production and the potential for contamination with human and animal pathogens. Furthermore, mice-derived mAbs may result in an immunogenic reaction in humans; a reaction which may be avoided by using mAbs derived from plants.

Plants have the capability to churn out large amounts of mAbs, with low production costs, the ability to be rapidly scaled up to meet market demand and reduced risk of contamination with human and animal pathogens. Systems for expressing mAbs have been extensively tested to permits in tobacco, but cereal grains are now being used because protein accumulation in dry seeds allows long-term storage at ambient temperatures with little degradation or loss of activity. Other crop plants that are used to produce antibodies include potatoes, alfalfa and rice.

Vaccines

A vaccine is an antigenic substance — either a protein, peptide, attenuated (weakened) living organism or a dead organism — which is able to draw forth an immune response that safeguards against development of an infectious disease. Vaccines are designed to 'educate' the immune system, so that upon exposure to the infectious agent it will already have the antibodies that will enable it to combat the infection.

Considerable research into the use of transgenic plants as an alternative to conventional vaccine production systems. PMV production appears to offer advantages in protein production and storage; and potential advantages in the distribution of vaccines to developing countries.

Other Therapeutics

Apart from antibodies and PMVs, scientists have also examined the potential for using plants to produce a wide assortment of other therapeutic agents, including hormones, enzymes, interleukins, interferons and human serum albumin. Proteins found in breast milk, such as lactoferrin, lysozyme and α-casein, have been expressed in plants. Addition of these therapeutic proteins to fortify other food products could aid in improving infant health.

Industrial Products

Plant molecular farming can also be beneficial making products with industrial applications including biofuels (biodiesel and ethanol), industrial oils, natural sweeteners, taste-modifying proteins, bioplastics, spider silk and elastic proteins.

Biofuels

Since biofuels currently cost more to produce than petroleum fuels, situation has generated considerable interest in developing new technologies to reduce the production costs of biofuels. Plant molecular farming is seen by many researchers to be one way of reducing these costs.

Sweeteners

Artificial sweeteners with a low calorific value (e.g. saccharin) have been developed as a substitute for sucrose and other natural sugars, especially for individuals who ail from diseases related to the consumption of sugar, such as diabetes, hyperlipemia, dental caries and obesity. However, some artificial sweeteners have unwelcome baking properties and aftertaste, which has led to research into alternatives. Potential alternatives to current artificial sweeteners that can be produced in plants include palatinose, brazzein, miraculin, isomaltulose and sorbitol.

Bioplastics

Bioplastics are considered an environmentally beneficial alternative to synthetic plastics due to their biodegradability and potential to decrease our reliance on petrochemical resources. The focus of current research into biodegradable plastics has largely been on polyhydroxyalkanoates (PHAs) —polyesters of hydroxy acids—that are produced by more than 100 different genera of bacteria. The monomers found in PHAs are highly diverse, suggesting that these plastics have a wide spectrum of physical properties and could be used as substitutes for polyethylene, polystyrene, polypropylene and PET. PHAs are also biocompatible, breaking down into molecules that are naturally inherent in animals, so they may have medical applications, for example, as implants, gauzes and suture filaments.

The cost of synthesizing of PHAs in GM or non-GM bacteria is high, with bacterial PHA currently priced five to ten times higher than petroleum-derived plastics. This constrains the use of bacterial PHAs to high-value products such as medical applications. There have been attempts to synthesise PHA in other organisms, including agricultural crop plants, to develop large scale production of biodegradable plastics economically. It has been calculated that bioplastic concentrations in plants would have to reach 15 per cent dry weight for commercial production to be economically feasible.

The polymer poly(3-hydroxybutyrate) (PHB) bacterium *Ralstonia metallidurans* from glucose is the most extensively studied PHA naturally synthesised by the. It has similar properties to polypropylene, although it is more brittle, making it less stress resistant for industrial applications.

It is naturally present at low levels in the cell walls of some plants. Higher levels of it were first produced in plants in 1992 in the model plant *Arabidopsis thaliana*. Since then researchers have attempted to produce PHB in a number of different plants with varying results; including oilseed rape, maize, sugarcane, flax, cotton, soyabean, palm oil, tobacco, switch grass, sugar beet, potato and alfalfa.

Spider Silk

Spiders spin silks with different properties for constructing webs, cocoons and draglines. For example, 'dragline' silk, used for the frames of a spider's web and for safety lines, is sturdier than high-tensile steel, but weighs less than one-tenth as much. Its tensile strength compares to that of the synthetic fibre, Kevlar, but it is more elastic. 'Capture spiral' silk has a lower tensile strength but can be extended to more than twice its length before breaking. These exceptional material properties have encouraged research into developing spider silks for industrial applications. Creating GM plants that can synthesise spider silk may allow large-scale production of these fibres.

Resilin

Resilin, an elastic protein found in specialised regions of the cuticle of most insects has low stiffness, high strain and efficient energy storage, and hence facilitates flight and jumping in insects. Its resilience and durability of could have both industrial and medical applications; for example, in spinal disc implants, vascular prostheses and high-efficiency rubber. There is an ongoing research on the possibility of using GM plants as a production system for resilin.

Broadacre Grain Crops and Oilseeds

Antibodies

The antibody 'T84.66' which recognises a marker for colorectal, lung, breast and pancreatic cancers, has been tested in Germany in both cancer imaging and therapy and has been produced in several different plant systems, including wheat, rice, tobacco and pea. An anti-Herpes Simplex Virus-2 (HSV-2) antibody, developed and expressed in soyabeans in the United States, has been shown to prevent vaginal HSV-2 transmission in mice. If a similar efficacy is shown in human clinical trials, then this antibody could represent an inexpensive protective measure for this disease.

Vaccines

A vaccine for the *E. coli* Heat-Labile Toxin protein diarrhoea caused by this toxin has been expressed in maize and potato in the United States to combat. The safety of the vaccine from each source is currently being tested in two independent phase I clinical trials.

Treatment of Allergic Reactions

Two proteins separated from Japanese cedar pollen (Cryj1 and Cryj2) have been expressed at a high level in GM rice in Japan. Initial testing in mice has shown that the consumption of this rice successfully treated allergic reactions to Japanese cedar pollen by causing immune system tolerance to develop. If this concept proves to work in people, treatments for a range of allergic reactions could be devised.

Research-grade Proteins

Aprotinin extracted from bovine lung tissue or derived from cows is an intravenously administered protein that helps prevent bleeding following cardiac surgery by blocking the action of certain enzymes in the bloodstream that dissolve blood clotsTP PT. It also has some other related research and manufacturing applications through maintaining protein stability. It has been successfully expressed in maize and tobacco plants.

Avidin is a protein naturally found in egg whites and has applications in research and diagnostics. The chicken avidin gene has been successfully expressed in maize. Avidin from GM maize was commercialised in the United States in 1997 and β-glucuronidase derived from bacteria is an easily visualised reporter gene that is used in plant research and diagnostics has been expressed in many plant species, including maize. Lactoferrin, an iron-binding protein found in milk and some white blood cells, can be used to treat gastrointestinal infections in addition to topical infections and inflammations. The lactoferrin gene from humans has been successfully expressed in rice and maize. Lactoferrin derived from rice for research purposes has been commercialised in the United States.

Lysozymes are enzymes that can destroy the cell walls of certain bacteria and have practical application as antimicrobial agents. A human version of lysozyme has been successfully integrated in GM rice plants. Lysozyme isolated from this GM rice has been commercialised in the United States.

Trypsin is an enzyme that cleaves peptides and proteins at specific places. The trypsin gene from cows has been successfully expressed in GM maize plants.

Treatment for Cystic Fibrosis

Gastric lipase may be of used to treat cystic fibrosis and pancreatitis. Researchers from the French company Meristem Therapeutics have successfully expressed gastric lipase in GM maize plants. The resulting drug is currently undergoing 'formulation optimisation' to optimise the delivery and activity of Merispase.

Cost-effective Biofuels

GM maize seeds are produced with increased sugar production. The modified seeds contain an enzyme that turns maize starch into sugar for ethanol production. This heat-stable enzyme should make ethanol easier to produce and reduce costs by eliminating the need for mills to add liquid enzymes.

Alternative Sweeteners

Brazzein with an intrinsic sweetness 500–2000 times that of sucrose is a protein. It can be isolated from the fruit of the African plant *Pentadiplandra brazzeana*, but it is uneconomical to produce on a commercial scale due to the limited availability of the fruit and complications associated with large-scale production of the native plant. Widespread commercial production of brazzein is, therefore, only likely to occur if the protein is expressed in an existing crop plant. GM maize plants expressing high levels of brazzein of up to 4 per cent of total soluble protein in maize seed were developed in the United States by ProdiGene, who reported brazzein accumulation. Tests indicated that maize germ flour containing brazzein could be directly used in food sweetening applications.

Production of Bioplastics

Moderate accumulation of PHB in commercial crop plants, with concentrations of 5.7 per cent dry weight and 7.7 per cent mature seed dry weight have been achieved for maize and rapeseed, respectively. Leaf chlorosis developed at higher levels of PHB accumulation in maize. No changes in size, appearance and germination frequency of GM rapeseed seeds were observed.

Tobacco Products

Tobacco produces abundant biomass, making it a useful platform crop for many plant molecular farming applications.

Antibodies

The antibody 'CaroRx™' binds specifically to *Streptococcus mutans* preventing the bacteria from stricking to the teeth. CaroRx™ produced in tobacco has reached phase II clinical trials in the United States.

A mAb recognising human chorionic gonadotropin is being expressed in tobacco in Germany. This antibody could potentially be used for pregnancy detection, emergency contraception and diagnosis or therapy of tumours that produce the pregnancy hormone human chorionic gonadotropin.

Antibodies have been developed that could be used to treat non-Hodgkin's Lymphoma. These have been expressed in a GM tobacco virus, which is used to infect tobacco plants, which then express the antibodies in the infected plant tissue. Researchers in Cuba have developed GM tobacco plants that make a mAb that is used for purifying a component of a vaccine for Hepatitis B.

Vaccines

An oral PMV for hepatitis B in humans has been successfully produced in tobacco plants, potato and lettuce.

A PMV to combat the infectious agent of bubonic plague has been developed in the United States. Researchers have modified tobacco plants to generate high levels of different antigens, from the plague bacterium *Yersinia pestis*. Early animal trials have shown that the vaccine elicits a protective immune response. Genetically modified tomato plants have also been used for plague vaccine proof-of-concept studies.

Alternative Sweeteners

Palatinose, a low calorie alternative to sucrose, has been produced industrially by using bacteria such as *Erwinia rhapontici* to convert sucrose into palatinose. Since this method is costly and the scale is limited, the widespread use of palatinose as an alternative sweetener is limited. The enzyme that converts sucrose to palatinose has been successfully cloned from *E. rhapontici* and expressed in both tobacco and potatoes in Germany. Tobacco though is not the ideal plant for this application whereas GM potatoes converted almost all their sucrose into palatinose, without affecting growth or development.

Spider Silk

GM plants that can synthesise spider silk have to be produced to allow large-scale production of these fibres. Successful expression of spider silk proteins has been achieved in tobacco and potato in Germany, with an accumulation of up to 2 per cent of total soluble proteins in leaves. Spider silk proteins have also been synthesised in *A. thaliana*. However, silk proteins isolated from these GM plants have not yet been successfully spun into a fibre. Given that silk properties are dependent on the structure of the fibre, which is influenced by the way the proteins are spun, advances in artificial spinning technologies are required before spider silk fibres will be commercially available.

Commercial Use of Transgenic Plants

The main goal of producing transgenic plants is to increase the productivity. In 1995-96, transgenic potato and cotton plants were used commercially for the first time in USA. By the year 1998-99, five other major transgenic crops cotton, maize, canola, soybean, and potato were introduced to the farmers. These accounted for about 75% of the total area planted by crops in USA. There are still a lot of concerns regarding the harmful environmental and hazardous health effects of transgenic plants. The major areas of public concern are- the development of resistance genes in insects, generation of a super weeds by mutation etc.

Bioethics in Plant genetic Engineering

There are issues and concerns regarding the use of transgenic crops and their effects on the health and the environment in general. The major concerns about GM crops and GM foods are:

a) Effect of GM crops on biodiversity and environment- As the GM crops are created artificially, there is no natural process of evolution in their development. Hence, there is a question of this affecting the biodiversity and overall effect on the environment.

b) The risk of transfer of transgene from GM crops to pathogenic microbes- Antibiotic marker genes are used to identify and select the modified cells. If GM food containing antibiotic resistance marker gene is consumed by animals and humans, there is a risk that the transgene will transfer from GM food to microflora of human and animals. This may lead to the gut microbes to become resistant to antibiotics.

c) The transfer of genes from animals into Gm crops for molecular farming may change the fundamental vegetable nature of plants.

d) The GM crops may bring about changes in evolutionary patterns. The plants adapt to the changing environment in the natural way by changing their genes and developing better races with superior traits which ultimately leads to the development of evolved races and varieties.

e) There is a risk of transferring allergens (usually glycoproteins) from GM food to human and animals.

f) There is a risk of "gene pollution" i.e. transfer of transgene of GM crop through pollen grains to related plant species and development of super weeds.

g) There are also some religious issues related to the consumption of transgenic plants with animal genes introduced into them, especially, for some strict vegetarian people and some ethnic groups with certain food preferences and restrictions.

h) There is a need to study thoroughly as to how the genetically engineered plants will affect the ecological balance, once they are released in the environment.

References

Bhat, S. F., and S. Srinivasan, "Molecular and genetic analyses of transgenic plants: considerations and approaches", *Plant Science,* 2002.

Domach, M. M. *Introduction to biomedical engineering*. Upper Saddle River: Pearson Prentice Hall. 2004.

Friedberg, E.C. *DNA Repair*. New York: WH Freeman and Company. 1985.

Jain, S. Mohan, D. S. Brar, and B. S. Ahloowalia, eds., *Molecular Techniques in Crop Improvement*, Dordrecht, Boston, and London: Kluwer Academic Publishers, 2002.

Twyman, Richard M. "Molecular farming in plants: host systems and expression technology". *Science Direct*. 28 October 2003.

6

Genetic Improvement of Orphan Crops

Orphan crops are also known as underutilized-, lost- or disadvantaged-crops. Most of these understudied crops are staple food crops in developing world. Some of the most important orphan crops belong to cereals [e.g., finger millet (Eleusine coracana) and tef (Eragrostis tef)], legumes [cowpea (Vigna unguiculata), and bambara groundnut (Vigna subterranea)], root crops [cassava (Manihot esculenta), and yam (Dioscorea sp.)], fruit crops [banana and plantain (Musa spp. L)], and many vegetables. These crops grow better than major crops such as maize and wheat under extreme environmental conditions. However, due to lack of genetic improvement, these crops produce inferior yield in terms of both quality and quantity. The major bottlenecks affecting the productivity of orphan crops are genetic traits such as low yield (for example, in finger millet and tef), poor in nutrition [cassava and enset (Ensete ventricosum)], and production of toxic substances [cassava and grass pea (Lathyrus sativus)].

Modern crop breeding utilizes techniques such as Marker-Assisted Breeding, and reverse genetics approach known as TILLING (Targeting Induced Local Lesions IN Genome). The application of these techniques to the understudied crops is important in order to boost productivity and feed the largely underfed and undernourished population of Africa. TILLING is a recently developed improvement method but proved to efficient in the detection useful mutations in crops such as wheat, barley, maize, rice and sorghum.

Orphan crops are not produced widely around the world, they are not traded to any significant extent in international markets, and they receive considerably less attention than the major crops from international or regional crop experiment organisations. Nevertheless, orphan crops are valued culturally, often adapted to harsh environments, nutritious, and diverse in terms of their genetic, agroclimatic, and economic niches.

Important new opportunities for improving orphan crops now exist in the tools gained through experiment on major crops and on model species, notably *Arabidopsis*. Recent achievements in the fields of genetics and genomics provide a more unified understanding of the biology of plants, which in turn can provide new opportunities for applying advanced science to orphan crops. A large discrepancy exists between the potential role of orphan crops in improving food security and the small amount of attention they have received.

Justifying significant investments orphan crop improvement requires a shift in investment from individual crops to whole sets of crops with common genetic structures. This process is identifying the correct balance of scientific investments between major and orphan crop development in poor countries so that the spillover effects from major crop are maximised and the scientific foundations for orphan crop are ensured. At the same time, attention must be paid to conventional breeding and delivery in both major and orphan crops. Crop improvement entails far more than biotechnology, and biotechnology entails far more than transgenic applications of Bt and herbicide resistance, which have been most commonly reviewed in the policy literature.

Description of Orphan Crops

Different names are used interchangeably to describe the range of orphan crops. Some of these names are, underutilized crops, lost crops of Africa, minor crops, neglected crops, and crops for the future.

Three criteria must meet in order for the plant to be considered underutilized or orphan crop,

i) proven food or energy value,

ii) the plant has been widely cultivated in the past, or the plant is currently cultivated, in a limited geographical area, and

iii) currently cultivated less than other comparable plants.

According to Naylor *et al.* twenty-seven orphan crops within developing countries are annually grown on about 250 million ha of land.

Although orphan crops are many in number, brief description is given below for the most important ones interms of the area they are grown and/ or population they feed. These include cereals (e.g. millet, tef, fonio), legumes (cowpea, bambara groundnut, grass pea), and root crops (cassava, yam, enset).

Finger millet (*Eleusine coracana*) is the most important small millet in the tropics and is cultivated in more than 25 countries in Africa and Asia predominantly as a staple food grain. The plant is tolerant to drought. The seed of finger millet contains valuable amino acid called methionine, which is lacking in the diets of hundreds of millions of the poor who live on starchy staples such as cassava. Finger millet is also a popular food among diabetic patients because of its slow digestion.

Tef (*Eragrostis tef)* is grown annually on over 2.5 million hectares of land mainly in Ethiopia. The plant is tolerant to abiotic stresses especially to poorly drained soils where other crops such as maize and wheat could not withstand. In addition, the seeds of tef produce healthy food because they do not contain gluten for which large portion of the population are allergic. Unlike other cereals, the seeds of tef can be stored easily without losing viability under local storage conditions, since it is not attacked by storage pests.

Fonio (Acha, *Digitaria exilis* and *Digitaria iburua*) is an indigenous West African crop. It is grown mainly on small farms for home consumption. Fonio is not only tolerant to drought but also a very fast maturing crop. It is also nutritious because it is rich in methionine and cystine, the two amino acids vital to human health and deficient in major cereals such as wheat, rice and maize.

Cowpea (*Vigna unguiculata*) is a leguminous crop annually grown on about 10 million hectares of land mainly in Africa. The crop is tolerant to drought and heat. It also performs better than many other crops on sandy soils with low level of organic matter and phosphorus. Since cowpea has quick growth and rapid ground cover, it is a useful crop in controlling erosion.

Bambara groundnut (*Vigna subterranea*) is an annual legume crop grown for human consumption. The seeds of bambara groundnut are known as a complete food because they contain sufficient quantities of protein, carbohydrate and fat. The average composition of the seed is 63 percent carbohydrate, 19 percent protein, and 6.5 percent oil.

Grass pea (*Lathyrus sativus*) is another leguminous plant commonly grown for human consumption in Asia and Africa. The plant is extremely tolerant to drought and is considered as an insurance crop since it produces reliable yields when all other crops fail. Like other grain legumes grass pea is a source of protein particularly for resource poor farmers and consumers.

Cassava (manioc; *Manihot esculenta)* is staple food for about a billion people. The plant is tolerant to drought and also performs better than other crops on soils with poor nutrients. The major problems related to cassava are its very low protein content and the roots contain poisonous compounds called cyanogenic glycosides (CG) which liberate cyanide. Konzo is a paralytic disease associated with consumption of insufficiently processed cassava.

Yam (*Dioscorea sp*) represents different species under genera Dioscorea. It is grown on about 5 million hectares of land world-wide and staple food in west Africa. The roots are the edible part and looks like sweet potato (*Ipomoea batatas*) although they are not taxonomically related.

Enset (*Ensete ventricosum*) is commonly known as 'false banana' for its close resemblance to the domesticated banana plant. Unlike banana where the fruit is consumed, in enset the pseudostem and the underground corm are the edible parts. Enset is the major food for over 10 million people in densely populated regions of Ethiopia. The plant is considered as an extremely drought tolerant and adapts to different soil types. Since enset flour is rich in starch but not in other essential nutrients enset-based diets need heavy supplementation.

Role of Orphan Crops in Socio-economic conditions

Orphan crops play particular role in food security, nutrition, and income generation to resource-poor farmers and consumers in developing countries. These crops perform better than major crops of the world under extreme soil and climatic conditions prevalent in developing world particularly in Africa. Most of African orphan crops including finger millet and bambara groundnut are extremely drought tolerant while some others withstand water-logging for longer period than the major crops of the world.

In general, orphan crops are extensively grown in Africa. The total global production of three orphan crops, namely bambara groundnut, fonio and yam comes from Africa. Africa also devotes large area of land for cassava, millet, plantain and taro cultivation. However, the total acreage and total production are not comparable for last four crops in Africa. For example, regarding cassava, Africa contributes for about 65 percent of the global area but produces only 50 percent of the total world production. This might be due to the use of unimproved planting materials and poor management practices. Except for banana, the production of three other crops (cassava, millet and yam) has steadily increased over time in Africa. Yam is exclusively grown in Africa and the production of this crop has tripled in the last 25 years.

Orphan crops are also compatible to the agro-ecology and socio-economic conditions of the continent. However, when these crops are replaced by other newer crops for the locality, some problems were reported. The best example is from the study made in the Northwestern Ethiopia where the incidence of malaria has been elevated in the years when the cultivation of exotic crops specifically maize was increased at the expense of indigenous or orphan crops.

Malaria is the major health problem in the world particularly in Africa. In the year 2006, there were an estimated 247 million malaria cases causing nearly a million deaths, mostly of children under 5 years. The study by McCann and colleagues indicated that the pollen from maize facilitates optimum conditions for mosquito breeding. Mosquitoes carry *Plasmodium* parasites, the causal agent for malaria. Larvae of the mosquito had a survival rate of 93 percent when it fed on maize pollen, as opposed to a survival rate of about 13 percent when it fed on other possible food sources. As a result, the cumulative incidence of malaria in high maize cultivation areas was 9.5 times higher than in areas with less maize.

Limitations of Orphan Crops

Although orphan crops perform better than major crop under extreme environmental conditions and fit to the socio-economic conditions of the developing countries especially in Africa, they have also a number of limitations. The major bottleneck is related to the little genetic investigation made on these crops. Almost all orphan crops are studied by poorly funded researchers based in the developing nations where resources for conducting research are limiting. The majority of these researchers have little chance to establish partnerships with the scientific community especially with those in the developed countries. Some of the outstanding bottlenecks related to orphan crops are indicated below:

— Poor grain yield: most orphan crops particularly cereals such as tef, millet and fonio produce extremely low seed yield.

— Poor in nutrient content: although root and tuber crops such as cassava and enset produce high yield, the products are largely starchy materials that are deficient in other essential nutrients particularly in protein. Although these crops are staple food crops for large number of Africans, supplementation with other nutrients is required.

— Unfavorable agronomic characters: Some of the negative features associated with the African rice (*Oryza glaberrima)* unlike the Asian rice (O. *sativa)* are rapid shattering of the seeds, difficulty of milling the grain, and lower seed yield.

— Abiotic stresses: Since most fertile lands are used to grow other crops than the orphan crops, the productivity of orphan crops under the less fertile and moisture deficit soils is extremely low.

— Hazardous or toxic products: The following orphan crops produce a variety of toxic substances that affect the health of human.

a. Cassava. The roots of cassava contain poisonous compounds called cyanogenic glycosides (CG) which liberate cyanide. Konzo is a paralytic disease associated with consumption of insufficiently processed cassava.

b. Hyacinth bean *(Lablab purpureus).* The pods and seed of hyacinth bean can be poisonous due to high concentrations of cyanogenic glycosides and can only be eaten after prolonged boiling.

c. African yam bean *(Sphenostylis stenocarpa).* The seeds of African yam bean contain anti-nutritional factors such as cyanogenic glycosides and trypsin inhibitors. Cooking is required to reduce toxins to safe levels, though this also decreases the level of nutrients in seed.

d. Grass pea. the seeds of grass pea contain a neuron-toxic substance called ODAP [â-N-Oxalyl-L-á, â-diaminopropanoic acid. ODAP is the cause of the disease known as neurolathyrism, a neurodegenerative disease that causes paralysis of the lower body. Serious neurolathyrism epidemics have been reported during famines when grass pea is the only food source.

Orphan Crops in a Development Context

Population growth over the next 30 years will be concentrated almost exclusively in the developing countries, where more than 1 billion people currently live on less than US$ 1 per day, more than 800 million people are undernourished, and 200 million children are underweight. This poverty is worst in rural areas where agriculture is the leading source of incomes and employment. The world's poorest regions are typically those where agricultural investments by the public and private sectors are extremely low. Unless some mechanisms can be found to stimulate agriculture, the outlook for these poor societies is bleak.

The role of agriculture in food security extends far beyond growth in crop yields and total production. Agriculture promotes food security primarily when it contributes to incomes and productive employment. Enhancing food security in the poorest regions requires investments in scientific study and training and the transfer of knowledge and technology to ensure wise management of resources and sustained capacity for growth. Moreover, food security dictates a focus on poor people's crops: subsistence and marketed crops grown in marginal areas where the poorest segments of the rural population are concentrated.

Few tabulations of the importance of orphan crops exist. Wheat, rice, maize and soybeans each occupy more than 70 million ha globally per year. Collectively, they cover 580 million ha and generate on the order of US$ 300 billion in gross value annually. Approximately two thirds of their combined area is within developing countries. Remarkably, these four crops also supply an average of about 1360 kcal of energy and 33 g of protein daily to individuals in poor counties.

Twenty-seven orphan crops within developing countries occupy areas of between 0.5 and 38 million ha. None is on the scale of wheat, rice, or maize, yet they total some 250 million ha. In addition, there are 70 million additional hectares planted to fruits and vegetables. Orphan crops take on even more significance within regions. Sorghum and millets are more important than rice and wheat, both in area and in contributions to the diet.

Similarly, roots and tubers play a dominant role, providing more than 400 kcal of energy per person per day. Given that 38% of Sub-Saharan Africa's total population is undernourished and that the number of undernourished children in that region is forecast to increase by 39% by 2020, the issue takes on added urgency. Important questions remain regarding the types of investments that are needed for individual crops or crop groupings.

Minimum investments in biotechnology approaches for individual orphan crops are likely to be at least as large as those for major crops for a particular trait. As a result, the economic incentives to invest initially in biotechnology for major crops seem clear. Once such investments have been made, however, incremental investments that extend the science to orphan crops could be extremely beneficial to some of the world's poorest populations, an opportunity often ignored.

Genetic Improvement of Orphan Crops

Crop production could be increased by either expanding the arable area or through intensification, i.e., using improved seed, fertilizer, fungicides, herbicides, irrigation, etc. According to Food and Agriculture Organization, agricultural intensification represents about 80 percent of future increases in crop production in developing countries. Based on this goal, crop breeders are focusing towards achieving improved cultivars that produce higher yields and at the same time tolerate to the sub-optimal soil and climatic conditions.

Among plant characters or traits that contributed for higher productivity in the last century, those which alter the architecture of the plant rank first. Architectural changes include alteration in branching pattern and reduction in plant height. The major achievement of Green Revolution in 1960's was due to the introduction of semi-dwarf crop varieties of wheat and rice along with proper crop production packages. These broadly adapted semi-dwarf cultivars were responding to fertilizer application; which led to tremendous increase in productivity. Currently, a number of genes affecting plant height are identified from major cereal crops including wheat, rice and maize. According to the International Food Policy Research Institute, Green Revolution represented the successful adaptation and transfer of scientific revolution in agriculture. However, since this agricultural revolution did not occur in Africa, crop productivity remains very low.

Modern improvement techniques are not yet employed in orphan crops. Breeders of orphan crops are mostly dependent on the conventional techniques such as selection and hybridization. Only limited numbers of breeders implement modern techniques such as marker-assisted breeding and transgenics. Genomic information such as whole-genome sequencing are not yet available for orphan crops. In order to feed the ever-increasing population of Africa, agricultural revolution is needed to boost productivity of orphan crops through the implementation of modern technologies proved to be effective for major crops of the world.

Molecular Tools for Germplasm Improvement

Over the past few decades, large investments have been made in biotechnologies for the major crops in industrialised countries. Molecular approaches have been widely used to describe and manipulate crop genomes and to better understand the genes and biochemical pathways that govern traits such as yield, disease and pest resistance, abiotic stress tolerance, plant architecture, and quality.

Many national agricultural research programs (NARs) in developing countries have also made some investment in molecular technologies as applied to the major crops, such as rice and maize, but the extent of that investment varies sharply across countries. Nations with strong NARs are generally those with strong economies, such as China, India, and Brazil. Even in those countries, however, biotechnology investments are rarely applied to orphan crops.

More than a decade of investment has developed powerful technologies for major crops, making genetic information for crop improvement more accessible. With this technology, DNA-based approaches have been applied in two broad areas of molecular breeding. The first, marker-assisted breeding, is aimed at enhancing the power of conventional genetic analysis and manipulation. The second is transgenics.

The first step of molecular breeding uses molecular markers as tools to detect the extent and structure of genetic variation, providing insights into the diversity of crop varieties and potential contributions offered by their wild relatives. Diversity seen at this level can be used to inform the selection of parents with a high likelihood of producing novel progeny in a crossing program.

Molecular tools are also used in the analysis of inheritance of key crop traits, including those that are subject to complex inheritance due to the involvement of numerous genes. Once the genetic basis of a trait is well understood, molecular tools can be used to select for specific desired genes or combinations of genes.

The second avenue involves the transfer of genes from one genotype to another. This approach, often referred to as "genetic engineering", may utilise man-made genes, or natural alleles. Thus, truly novel traits can be added to a crop. The two broad areas of molecular breeding converge when genetic engineering is applied to enhance the efficiency with which native genes are moved within a gene pool.

Genome Analysis and Molecular Breeding

Evidence indicates that all plant genomes have a great deal in common in their gene content, biochemical pathways, and chromosome organisation. Different plant taxa have different versions of the same genes at a given position or locus in a genome, but the order of loci is conserved to varying degrees across even distantly related crops, a phenomenon known as synteny. Because plant genomes are somewhat similar, investment in "model" species, chosen because they are easily studied, has paid off. Models have been selected for each of the most important plant groups, and research has been accelerated through large-scale projects.

The flowering plants can be divided into two main groups: monocots and dicots. The former includes the grasses, which in turn include the major cereal crops. The latter include the legumes, many roots and tubers, and vegetable crops. The genome of the model dicot species *Arabidopsis thaliana* has been fully sequenced, and a great deal of basic research has been done on this species. More recently, the draft genomic sequence of the rice (*Oryza sativa*) genome was made available, which is useful both for rice and its monocot relatives.

Ninety-eight percent of the proteins found in maize, wheat and barley are also found in rice. Although the progenitors of rice and *Arabidopsis* diverged 150 to 200 million years ago, more than 80% of the genes that have been documented in *Arabidopsis* have also been found to have related genes in rice. The genes involved in many biochemical pathways and processes are very similar across the plant kingdom. Functions such as gene regulation, general metabolism, nutrient acquisition, disease resistance, general defense, flowering time and flower development are largely conserved across taxa.

Comparative mapping studies have revealed that the genomes of plant species within families are syntenous—that gene order is conserved for chromosomal segments. Given the similarities among crop genomes, particularly among plant species within a family, it seems possible that research on major crops or model species would benefit a substantial number of related crop species in the same families. The potential spillover benefits are likely to differ, however, across taxa and region depending on the particular biology of the species, the constraints faced by farmers, and seed systems.

Genetic Diversity

The use of molecular techniques for analysis of genetic diversity and the structure of germplasm, the first conceptual step in marker-assisted breeding, has been fruitful for many species. This type of research can lead to better conservation of crop genetic diversity, more efficient selection of accessions for phenotypic characterisation, and the identification of useful variants for conventional plant breeding.

Some molecular techniques (e.g., amplified fragment length polymorphism, or AFLP) can be applied to any species, without the need for crop-specific DNA sequence information. Such analysis can be conducted in any laboratory for which minimum levels of basic training, supplies and equipment are available. Even with the most seemingly straightforward techniques, however, it is difficult to achieve smooth and productive operations under the conditions encountered by many scientists in developing countries. Administrators often err in thinking that investment in facilities alone is sufficient for scientific success, and they neglect the substantial costs incurred for training, consumables, and the diversion of key personnel from field-based efforts. Even when researchers are able to produce good-quality molecular data, they may lack access to expertise in some of the analytical procedures needed to interpret the results.

This area of research provides a clear example of one in which spillovers from investments in facilities, human capital, and supplies for major crops could be quite large for orphan crops. Once researchers are well trained in the techniques and have the necessary equipment, they could apply their skills to a wider range of crops, or they could help to train others to analyse genetic diversity of orphan crops.

Marker-assisted Selection (MAS)

Marker assisted selection (MAS) is the identification of DNA sequences located near genes that can be tracked to breed for traits that are difficult to observe. MAS can be relatively straightforward for genes conditioning large phenotypic effects, such as strong effects on disease resistance, and its use for germplasm improvement is beginning to prove successful for some types of field applications for major crops. Two new rice varieties with bacterial blight resistance were introduced commercially in Indonesia in 2002.

Rice lines carrying multiple resistance genes are being produced by several national programs following technical investments at the International Rice Research Institute (IRRI). Strong national programs, such as those in China, are effectively using MAS in some cases, e.g., improvement of quality traits in rice and improvement of fiber strength in cotton.

In "single large-scale MAS", selected genes are fixed by marker-assisted selection, and the constitution of the rest of the genome is determined by conventional field-based breeding. This approach has been used to develop virus resistance in maize, rice, and beans. In recent cost analyses of marker-assisted versus conventional selection, the cost of MAS for a recessive mutant gene associated with quality protein maize was found to compare favorably with the conventional screen.

As basic experiments on key plant traits advances and as marker technologies continue to become increasingly simple and powerful, more applications of MAS are anticipated with the potential benefit of faster varietal release. Morris et al. showed positive cost-benefit results of MAS, and there are many similar unpublished examples from the private sector. In the published literature the cost-benefit results of MAS are mixed, and many publications only report work aimed at eventual application of MAS as opposed to field-level use of MAS for practical crop improvement.

Due to the relatively high cost of MAS compared to conventional field screening, MAS is likely to be practical for orphan crops in the poorest countries only if partnerships with well-funded research groups allow free use of existing technologies. It is important, therefore, to assess various traits and trait categories for which there has been some proof-of-concept in non-orphans for MAS before considering what would be required to move the benefits to a broader set of plant species.

In viewing cases in which genetic research has allowed for progress in MAS, it is important to distinguish between two types of traits: those conditioned by single genes with large effects showing simple inheritance (qualitative traits), and those showing complex inheritance (quantitative traits). Important plant traits that frequently show simple inheritance include some resistances to diseases and insects and some important characters that relate to plant growth and development. It is relatively easy to locate these genes with precision, and in many cases such genes have been cloned. It is often straightforward to select for these genes in a conventional breeding program, in which case MAS does not necessarily offer substantial benefits.

There are cases, however, in which markers are particularly beneficial; for example, when a disease resistance gene must be identified but the pathogen is not present at the breeding site, or when a gene is to be selected at the seedling stage for a trait that is only expressed later in development.

Most important traits are governed by multiple genes, each having relatively small effects. These "quantitative traits" have been difficult to understand and to manipulate in conventional crop breeding programs. The term QTL, quantitative trait locus or loci, refers to the chromosomal regions of genes that control quantitative traits. Molecular genetic approaches began to illuminate the genetic architecture of quantitative traits. By now, the chromosomal segments associated with many traits have been identified in a large number of experiments. As estimated by Goff et al., about two thousand cereal QTLs have been mapped.

While so-called anonymous QTL-associated markers are sometimes used for marker-assisted selection in crop improvement, the utility of such markers can be limited by a high degree of imprecision in mapping desirable loci of small effects. The identification of "candidate genes" makes it possible to localise desirable variants much more precisely.

Candidate genes are genes known or suspected to be involved in conditioning the phenotype of interest, such as disease resistance. The inference that a particular gene contributes to a given trait may be strengthened if it maps to a chromosomal region that has been associated with the trait in QTL mapping experiments.

Credible candidate genes co-localising with QTLs have been identified for several traits, including quantitative disease resistance in rice, wheat, bean, and potato. A number of experiments approaches have converged to allow some genes underlying QTLs to be cloned, and to set the stage for future QTL cloning in others. This process permits both the identification of potentially useful variants of agronomically important genes and the precise selection of those alleles found to be most useful.

One of the approaches that puts QTL cloning within reach is the increasing availability of express sequence tags (ESTs), short segments of sequenced gene transcripts that provide information on genetic expression, function, and heritability. Tens or hundreds of thousands of ESTs are available for major crop species, but the number of sequences for orphan crops is quite meager. The combination of QTL mapping data and mapped EST data as a powerful tool for determining the genes that underlie quantitative traits.

Sequence data on expressed genes and on plant and crop genomes are rapidly accumulating and present powerful tools for plant science, provided that scientists are able to access and exploit the data efficiently. The sequence data sets are, in themselves, imposing and potentially cumbersome. But with the help of increasingly powerful and friendly databases, biologists and breeders can now gain access to genetic information that allows them to identify and exploit natural variation for a wide variety of crops in ways that were previously not possible.

Regulatory Genes

The genetic complexity underlying quantitative traits often makes it difficult to seek trait-marker correlations that can be used for indirect selection. For instance, a great number of genes (hundreds or thousands) are involved in plant defense against insects and pathogens. This complexity may seem defeating, but it may be possible to improve complex traits by manipulating the genes that sit at the top of the regulatory hierarchy and control large numbers of other genes. In fact, there is growing evidence that many QTLs are explained by transcription factors-genes that control the expression of other genes. The identification and utilisation of the "best" alleles of these genes is a promising avenue of research, both via MAS and via transgenic approaches. Transformation experiments enhancing the stress-responsive expression of a transcription factor that controls drought-related genes led to increased tolerance to drought, salt stress, and freezing in *Arabidopsis*.

Similarly, overexpression of a regulatory factor involved in plant defense in transgenic *Arabidopsis* led to enhanced resistance to diverse pathogens. Some of the genes involved in regulation of the defense response are also implicated in responses to other stresses. The identification of these transcription factors seems likely to play a useful role in genetic improvement for a wide range of food crops.

The use of direct gene transfer, the second tool for "molecular breeding" of crop traits, employs recombinant DNA technologies to insert one or more genes into the crop plant's genome. While most transgenic research and applications have focused on a few major crops to date, the potential to extend applications of this technology to orphan crop improvement may be significant. Promising transgenic lines for about 20 different crops, including pepper, squash, and sweet potato, are currently being field tested by public research institutions in at least 10 developing countries. These lines, which have been engineered for particular traits like virus resistance and pest control, could provide significant benefits to poor farmers who cannot afford chemical controls for crop loss.

A wide range of approaches employing transgenes is available for controlling herbivory, but by far the most extensively used method to date has been "Bt" technology. Different Bt toxins are lethal to the immature or larval, herbivorous stages of lepidoptera (adults are the butterflies and moths), diptera (the flies, including mosquitoes), and coleoptera (the beetles, including weevils), many of which are major crop pests.

Beyond the use of Bt, several additional insecticidal products are produced by plants, animals, and microbes that are also being evaluated for the control of herbivory. For example, many plants store large amounts of specialised proteins in their seeds (called storage proteins), and some of these proteins have insecticidal properties.

Insects that consume plant seeds are particularly dependent on their ability to digest seed proteins using one or more proteases, and these inhibitors have been engineered into other plants to interfere with the feeding of insects on leaves and flowers. Protease inhibitors have also found to be effective against some of the plant pathogenic nematodes.

A number of enzymes have also been tested for their possible use as insect resistance determinants. One of these enzyme groups is the chitinases, which target an essential structure of macromolecules (chitin) in insects and fungi. Three additional groups of enzymes (conifer monoterpene synthases, bacterial cytokinin biosynthesis enzymes, and cholesterol oxidase) have been designed to modify the chemistry of the plant in ways that make it less palatable as an insect food source.

Finally, in many crops, especially the grains, a major product stored in seeds is starch. Insect pests of these crops depend on their ability to digest starch using the enzymes α and β-amylases. Some plants encode a gene for an inhibitor of α- and β-amylase; these genes have been used in a variety of cases to produce insect resistant plants.

Transgenic approaches for reducing crop losses from disease have also been widely explored, with significant success in the area of plant viruses. Reports of success on transgenic technologies for controlling plant viruses abound in the literature and encompass numerous crops including rice, wheat, maize, barley, perennial rye, legumes, potato, sweet potato, several temperate vegetable species, and papaya.

Plant viruses represent a major biotic constraint on agricultural production; viral symptoms commonly include stunting, reduced yields, or death. Viruses are usually transmitted to plants via insect, beetle, nematode, or other types of vectors, and as a result, resistance is often the only effective form of viral control in agricultural systems. While genetic resistance exists in many crops, it is not always well understood or identified. For orphan crops in particular, even the most basic knowledge of resources for genetic resistance is typically lacking.

The mid-1980s confirmed that when portions of plant viral genomes, such as the viral coat protein gene, were transferred and expressed in a plant genome, resistance to the virus was observed. Many types of sequences from both RNA and DNA plant viruses have been expressed in a diverse set of crops resulting in resistance to a wide array of pathogens. More recently, scientists have turned their attention to resistance resulting from expression of genes from other sources, including mammalian antibodies, anti-viral compounds, and sequences that act to silence viral genes or silence host genes necessary for viral infection. To date, however, resistant varieties resulting from "second-generation" strategies (i.e., those involving gene silencing or expression of non-viral sequences) are not yet in use.

Nonetheless, future strategies to limit crop losses to viral disease will most likely encompass upstream research on sequences that silence plant genes and on direct gene transfer to move plant genes beyond the boundaries defined by sexual hybridization.

Genetic engineering for insect or virus resistances is a so-called "implementation technology", meaning that once the genes are available, they must be combined using an "enabling technology" required to move the genes into target crops. Thus, in principle the sequences used that encode the "pesticidal" activity, combined with appropriate regulatory sequences that work in the target crop, can be used in any crop plant. In this way, genes isolated originally for use in a major crop such as wheat or soybeans can be readily modified for use in orphan crops. The technical challenges for scientists applying these technologies to orphan crops are threefold.

The first challenge relates to the extent of knowledge about the sensitivity of the pathogens or pests of orphan crops to forms of control that have been useful in developed countries with temperate environments. Work in India on insect resistance for chickpea showed that the proteins encoded by the previously isolated proteinase inhibitor genes were at best marginally effective against insect pests of chickpea. The researchers then turned to isolation of orthologous genes from crop plants such as Asian winged bean to find more effective proteinase inhibitors.

The second challenge concerns the availability of techniques for transferring genes into orphan crops. In some cases, the transferability of gene transfer methods has been fairly easy and direct from model species to orphan crops. But in other cases, significant new and expensive work has been needed to modify-or to develop entirely new-transformation technologies for these crops.

And finally, a major limitation of the use of these promising technologies in the orphan crops to date is the paucity of approaches that can be successfully applied to the control of diseases caused by bacteria, fungi, the oomycetes, and nematodes.

Genomics-guided Transgenes (GGT)

Genomics approaches and transgenic technology converge when natural allelic variation is utilised through direct gene transfer. Useful alleles can be transferred from a wild relative of a crop by conventional breeding, but it may be difficult or even impossible to recover the desired plant type in a reasonable period of time. Using "genomics-guided transgenes" (GGT), native genes or homologous genes from closely related species that modify a plant's metabolism in a manner similar to natural or induced mutations are used for direct gene transfer. Dominant alleles that confer important agronomic traits of interest (e.g., pest and disease control, drought tolerance) but that are scarce in breeding populations can be identified and inserted into breeding lines, based on knowledge gained from model gene studies. An example of this approach is the use of the *Xa21* gene, which provides resistance to the bacterial blight disease of rice. *Xa21* was originally transferred from a wild relative to cultivated rice by conventional breeding.

The gene has been moved among genotypes by MAS and has also been cloned and utilised through genetic engineering. GGT can comprise a wide diversity of genes, although many of the modified traits using this approach to date were already familiar, such as male sterility, seedless fruits, delayed spoilage, and dwarf stature.

Given the range of molecular tools, what are the promising applications for orphan crops? The answers can best be explored when related major crops have received substantial research investment. The most obvious example is the grass family, or Gramineae, the most economically important plant family. There are around 10,000 graminaceous species, which include the world's most important cereal crops. Knowledge of the taxonomic structure facilitates the use of information from the more studied species to the orphans.

There are five sub-families in the grass family. Rice is a member of the ehrhartoid sub-family and has the smallest genome. The pooids include wheat, barley and rye. The panicoids include maize, sorghum, and pearl millet, all of which are important staples in sub-Saharan Africa and other areas. The chloridoids include the Ethiopian staple tef, as well as finger millet. The bambusoids include bamboo.

Within the grass family, genomes vary substantially. While the genomes show similarities in gene organisation, there is evidence of significant variation in genes, chromosomes and genomes. Based on a framework of restriction fragment polymorphism (RFLP) maps, the cereal chromosomes show gross synteny. At a finer level, however, there are many deviations from synteny, such that the predictive power between taxa ranges generally between 50 and 75%. Even between maize lines, the gene content of a given chromosomal region vary.

Some of the genes affecting important traits such as seed size, seed dispersal, and flowering time have been shown to exist in corresponding sites in the genomes of rice, sorghum and maize. Several QTLs conditioning resistance to the blast disease pathogen were found to exist in the same locations in both rice and barley.

In many cases, therefore, knowing the location of a gene or group of genes affecting a key trait in one species will be useful for efficient identification of genes that are of interest for another trait. If the actual genes are known, then the gene discovery process is less dependent on synteny, as the genes can be traced by sequence similarity. Similar examples are found for a range of plant families. For instance, the Solanaceae include many vegetable species that are extremely important as food and cash crops. For tomato, a substantial amount of genetic and genomic information is now available. The situation for potato and pepper is intermediate.

The chromosomes of these species are syntenous, and the genes for various characters have been found to reside at syntenous locations. For various African indigenous vegetables (AIVs) in the Solanaceae, there is virtually no information available, but information from the better-studied solanaceous crops would doubtless be an asset if an investment were made in the AIVs. A similar situation exists for the legume family. This mapping, while only partial, helps to identify niches of opportunity for biotechnology spillovers into orphan crops. Much greater challenges exist for orphan crops like quinoa and some roots and tubers that have no close model or major crop relatives.

Despite identifiable niches, the task of implementing biotechnology approaches to germplasm improvement is not simple, particularly in poor regions. The process of analysing the diversity of genes related to a trait of interest, and identifying and marking the useful alleles for selection, requires good access to current scientific literature and the internet, with at least modest bioinformatics capacity.

The best promise for implementing a specific biotechnology strategy for germplasm development in orphan crops holds when research on a particular trait is pursued across a group of related species, one or more of which is not an orphan. The most rapid progress is likely to be made when there is a model crop in the same family as the orphan crop.

In the sweet potato case, weevils are known to reduce yields by 60%, and in some cases by 100%. Average yields, currently at 4.4 tons/ha, would thus rise to over 7 tons/ha with fully effective resistance. Viewing potential success more conservatively, a 25% increase in yields from weevil resistance would raise the average to 5.5 tons/ha.

Given a sweet potato area of 570,000 ha in Uganda and a price of US$ 88/ton, the gain from achieving Bt-derived weevil resistance on half the area would be over US$ 27 million per year. If the technology could be distributed throughout the Sub-Saharan Africa on one-half of the sweet potato areas, the gross annual benefits would be about US$ 121 million. Twice the initial investment would be needed to move the resistant cultivars into the field in Uganda (US$ 10 million), and that five times the initial amount would be needed to disseminate the technology throughout the Sub-Saharan region (US$ 25 million). Even with these expenditures, the net payoffs would still be large. Moreover, because sweet potato is a vegetatively propagated crop, the biosafety risks of introducing Bt in this case would be much lower than for maize or cotton.

Investing in marker assisted selection for blast resistance in finger millet could also prove successful within the next decade, particularly given the gains made in identifying genes and QTLs for blast resistance in rice. Finger millet yields are typically reduced by 35% or more as result of neck and finger blast. Successful breeding for blast resistance could increase average yields from 1.3 ton/ha currently to over 1.75 ton/ha.

A more conservative estimate of potential success would be to achieve a 15% gain in yields from resistance (to 1.5 tons/ha on average) on one-half of the total area. Given an area of 3 million ha planted to finger millet in India, another 1 million ha in Africa, and a price of US$ 96/ton, the gross annual benefits would be more than US$ 38 million globally. Again, even if large costs were incurred to move the technology into the field, the net pay-offs would still be large.

Greater challenges exist with a crop like tef, where the technology path leading to the creation of dwarf varieties is more obscure. In all cases, however, there are benefits associated with the investments, including the enhancement of scientific capabilities and human capital development. The probability of success, the potential financial returns, the collateral benefits, and the constraints on technology adoption should all be weighed.

Success in the laboratory or experimental plot still needs to be integrated into a much broader scientific process in order to be successful in rural communities. A technological continuum exists for germplasm improvement, which ranges from simple selection techniques, to conventional breeding approaches with farmer participation, to biotechnology developments in advanced laboratories. Unfortunately, the continuing decline in funds for practical crop improvement, particularly in the international public sector, results in a fair amount of investment in research, and a weak application to practical problems. Applying molecular techniques to basic research in plant biology has led to a much deeper understanding of the genetics of key traits, but not always to a clear idea of how plants can be improved.

Appropriate use of biotechnology depends on the agricultural problem at hand, the biological properties of the crop, and the economic and social infrastructure that supports crop research. The best chances for harnessing the gains from biotechnology exist when the science is integrated into breeding efforts, farm management, and seed production and distribution. Even with integration, the benefits from advanced science depend critically on the institutional, human capital, economic, and political context of the recipient countries.

Useful generalisations about potential field-level applications of biotechnology for orphan crops are made difficult by the great diversity of crops and socioeconomic settings. Nevertheless, orphan crop improvement is likely to benefit from attention in three areas: dispelling the myth that biotechnology is equivalent only to genetic engineering applications of Bt insect resistance and herbicide resistance; creating new incentives for public-private partnerships; and fostering new institutional arrangements that combine efforts across whole sets of major and orphan crops.

Many people in the policy community—and in the general public—believe that agricultural biotechnology encompasses only transgenic approaches to insect and weed resistance. This focus on GMOs has pitted consumers against industry, Europe against the US, and environmentalists against each other as they debate the relative costs of pesticide use and genetic engineering. While the debates are valid in their own right, they overlook the fact that many new genetics and genomics techniques do not involve the insertion of foreign genes into plants.

The use of molecular tools to understand the genetic basis of crop traits for indirect selection and breeding is beginning to offer an alternative to traditional transgenics. These technologies will help alleviate some of the constraints surrounding intellectual property rights and biosafety that limit the dissemination of GMOs in developing countries. Many developing countries have been reticent to employ GMOs, lest their use interfere with trade and aid. It is the responsibility of agricultural policy analysts and practitioners to help redirect policy and public discussions on the future of biotechnology. The point should focus foremost on the needs of poor farmers and consumers, and the tradeoffs they face. To the extent that conventional breeding can be facilitated by advanced genetics techniques, biotechnology should be encouraged as one tool in the agricultural development package.

Public-Private Partnerships

If newer forms of biotechnology are to be applied, and especially those involving spillovers from one crop to another, partnerships will be necessary. Many partnerships exist and can be built upon, especially among universities and national agricultural research institutions (NARs). Two types of linkages that transcend the traditional university-NAR relationships are also likely to be key: honest broker organisations that bring together needs and skill sets, and the private sector that is the home for much of the relevant technology. The private sector is also essential in moving crops from the lab to the field. In many cases, the public sector has a comparative advantage in discovery, while the private sector has a comparative advantage in overcoming regulatory barriers and distributing seeds.

Creating new public-private partnerships will be essential for managing intellectual property issues. Ten companies, which own more than three-fourths of the agricultural patents in the US, are likely to be central in many of the biotechnology applications. Significantly, Dow, DuPont, Monsanto and Syngenta have already agreed to provide seed varieties, patent rights, and laboratory knowledge to African countries through the African Agricultural Technology Foundation (AATF).

The resulting spillovers into orphan crops could be especially important, although for this to happen, AATF will probably first need to succeed with a major crop such as maize. The private sector controls patents on transformation technologies for genetic manipulation as well as on the products themselves. In order for some of the newer forms of transgenics applications to work (i.e., those that do not involve Bt), patent rights on these transformation technologies will have to be relinquished not only for research, but also commercialisation.

The Center for the Application of Molecular Biology to International Agriculture (CAMBIA) has a major focus on securing patent rights for the public sector, particularly in developing countries. In addition, a consortium of US universities convened by The Rockefeller and McKnight Foundations has recently launched an initiative called the Public-Sector Intellectual Property Resource for Agriculture (PIPRA), designed to make agricultural technologies accessible for humanitarian uses in developing countries.

Intellectual property is not the only factor discouraging private and public partnerships. For much of Africa and parts of Asia and Latin America, the lack of biosafety protocols and their enforcement creates serious problems of legal liability. Without such protocols, external public and private organisations will be exceedingly reluctant to become involved with orphan (or major) crops, and for very good reasons. No responsible organisation wants to be a party to, for example, an "escaped" pathogen or an inserted gene that out-crosses into a weedy relative. Remy et al. say that transformed bananas resistant to two devastating diseases in Rwanda and Burundi have been developed in Belgium. Because of the lack of biosafety protocols, the transformations remain in the laboratory.

Issues of intellectual property rights and biosafety are not confined to the transfer of science to orphan crops; major crop improvement programs face similar problems in developing countries. From a policy perspective, new programs will need to be developed in the legal and safety areas-in many cases with the help of external aid and consultation-to facilitate the adoption of technology and its use in farmers' fields. In terms of policy priorities, the establishment of legal and safety measures is often more important that allocating additional funds to biotechnology for crop research, lest new technologies simply remain on the shelf. These constraints will likely be much lower when molecular markers and mapping techniques aid conventional crop breeding. Nonetheless, private sector cooperation may still be helpful in distributing seeds.

For major crops, such as wheat and rice, much of the linking of research with farmers has been performed by Centers of the CGIAR. Limited inroads on millets, cassava, bananas, and a few other crops have been made by additional CGIAR Centers, but there are large numbers of orphan crops where virtually nothing has been done. It is on this point that the new AATF instituted by the Rockefeller Foundation holds considerable promise. Its mission is to help design the relevant templates, protocols, and procedures that will lower the transaction costs of applying biotechnology to major and orphan crops in Africa. By doing so, it will provide a model for other regions.

The McKnight Foundation's Collaborative Crop Research Program is also aimed at enhancing the transfer of science from major to orphan crops and training scientists from poor countries in advanced genetics and genomics methods. Finally, the biotechnology programs being developed and promoted in Dutch, Swiss, and US aid agencies are contributing to progress in orphan crops. Special programs and new incentives within the scientific and development communities should be further encouraged to achieve widespread spillover benefits for poor farmers.

Applications of Modern Improvement Techniques

A number of molecular markers are implemented in modern plant breeding. These include Restriction Fragment Length Polymorphisms (RFLPs), Random Amplified Polymorphic DNAs (RAPDs), Amplified Fragment Length Polymorphisms (AFLPs) and microsatellites (Simple Sequence Repeats, SSR). Marker assisted selection (MAS) is the identification of DNA sequences located near genes that can be tracked to breed for traits that are difficult to observe. According to Collard and Mackill the following factors should be considered before selecting what type of DNA marker to be used in MAS: reliability; quantity and quality of DNA required; technical procedure for marker assay; level of polymorphism; and cost. Comparative mapping studies have revealed that the genomes of plant species within families are conserved for chromosomal regions. Hence, orthologous genes from orphan crops could be identified and isolated based on information from major crops.

Conventional breeding technologies including selection, hybridization and mutation breeding are all considered as non-transgenic methods. From modern techniques, marker-assisted breeding and TILLING are also non-transgenic. TILLING (Targeting InducedLocal Lesion IN Genomes) is a high-throughput and low cost method for the discovery of induced mutations. Transgenic and a modified form known as cisgenesis are widely applied to major crops such as rice and maize. Only few orphan crops have so far benefited from the techniques. Transgenic is considered as other advancement towards boosting crop yields and improving nutritional quality of crops. Due to high adoption rate, the global area under transgenic crops is tremendously increased from just 1.7 million ha in 1996 to about 134 million ha in 2009.

Future Perspectives and Recommendations

In order to boost productivity and diversify the food system in Africa orphan crops should be given due attention. The following points need to be considered in order to promote orphan crops research:

— Apply modern improvement techniques to orphan crops research. These include MAS, TILLING, tissue culture and transgenic techniques. The goals and techniques to be employed might vary for different orphan crops. However, the majority of research on orphan crops focuses on four areas, i) improving productivity per unit area, ii) breeding for tolerance against biotic and abiotic stresses, iii) enhancing nutritional quality through biofortification, and iv) removing toxic substances from some plant species.

— Financial and technical supports for researchers and institutions involved in orphan crops research since research on these crops are mostly under-funded and dependent on locally available meagre resources. These supports could be invested in training African scientists and developing infrastructure for African research institutes.

— Establish partnerships with public and private institutions. The partnerships could be made within and between research institutes, universities and private organizations.

— Create a network of orphan crops researchers at different levels: national, regional and international. The network will be an effective information exchange mechanism either among orphan crops researchers or with those working on major crops. This can be facilitated, for example, by forming internet discussion forum.

— Organize conferences, workshops or trainings related to orphan crops research and development. The First International Conference on African Orphan Crops was held in September 2007, in Bern, Switzerland in order, i) to address the major crop productivity problems related to orphan crops in Africa; ii) to propose the strategy of implementing modern techniques to orphan crops; and iii) to discuss the prospects and feasibility of modern crop biotechnology in African agriculture in general and orphan crops in particular through round-table discussions involving prominent scientists. About 80 researchers from four continents participated in the conference. The Proceedings of the conference has been recently published. The second conference on African orphan crops is expected to take place in Africa in near future.

Orphan- or Understudied-crops provide food for resource poor farmers and consumers in Africa. They also grow under extreme environmental conditions, many of them poorly suited to major crops of the world. Since Green Revolution did not occur in Africa, the continent did not benefit from the positive effects of this agricultural revolution that boosted the productivity of food crops in other parts of the world. The next Green Revolution for Africa needs to also include these locally adapted crops that are mostly known as orphan- or understudied-crops. Although these crops are largely unimproved, the implementation of modern improvement techniques on these crops has many advantages.

References

IPGRI. 2004. International Plant Genetic Resources Institute 2004). Promoting fonio production in West and Central Africa through germplasm management and improvement of post harvest technology. Final report.

James C. 2010. *Global Status of Commercialized biotech/GM Crops: 2009.* International Service for the Acquisition of Agri-biotech Applications (ISAAA) Briefs No. 41. Ithaca, NY.

Varshney RK, Close TJ, Singh NK, Hoisington DA, Cook DR. 2009. Orphan legume crops enter the genomics era! *Current Opinion in Plant Biology* 12:202–210.

Williams JT, Haq N. 2002. *Global research on underutilized crops.* An assessment of current activities and prospects for enhanced cooperation. ICUC, Souhampton, UK.

7

Risk Assessment of Transgenic Crops

Genetic engineering is an application of biotechnology involving the manipulation of DNA and the transfer of gene components between species in order to encourage replication of desired traits. Although there are many applications of genetic engineering in agriculture, the current focus of biotechnology is on developing herbicide tolerant crops and on pest and disease resistant crops. Transnational corporations such as Monsanto, DuPont, Norvartis, etc. which are the main proponents of biotechnology view transgenic crops as a way to reduce dependence on inputs such as pesticides and fertilizers. What is ironic is the fact that the biorevolution is being brought forward by the same interests that promoted the first wave of agrochemically-based agriculture, but this time, by equipping each crop with new "insecticidal genes," they are promising the world safer pesticides, reduction on chemically intensive farming and a more sustainable agriculture.

Worldwide, the area planted in genetically modified (GM) crops has increased dramatically in recent years. Between 1996 and 1999, it rose from 1.6×10^6 ha to more than 35×10^6 ha. This rapid increase has provoked an explosion of concern, particularly in Europe, over the health and environmental impacts of these crops. Despite claims of safety and warnings against popular panic, public concern over GM crops has resulted in changes in their marketing, labeling, planting, and trade. These changes have fueled an increasingly heated debate among environmental advocates, critics of industrial agriculture, seed companies, governments, and scientists.

GM organisms have the potential to both degrade and improve the functioning of agroecosystems. Depending on which GM crops are developed and how they are used, GM crops could lead to either increases or decreases in pesticide use, the enhancement or degradation of the ecological services provided by agroecosystems, or the loss or conservation of biodiversity. However, the current character of GM crop development provides cause for concern.

Biotechnology and Agriculture

Biotechnology is going to be an essential partner, if yield ceilings are to be raised, if crops are to be grown without excessive reliance on pesticides and herbicides, and if farmers on less favored lands are to be provided with crops that are resistant to drought and salinity, and that can make more efficient use of nitrogen and other nutrients.

Roughly 95% of the world's farmers live in developing countries. Most of these people engage in small-scale, community-based agriculture. Over long periods of time, these communities have constructed complex systems of knowledge about their environment. More recently, the "green revolution" succeeded in making food more easily available to most of the world's population. Increases in agricultural production were brought about by a combination of increased irrigation, more intensive use of fertilizers and plant protection chemicals, and the development of new crop varieties capable of responding to higher levels of inputs and management. However, this agricultural intensification often came at the expense of local ecosystems and human health. These changes reduced the ability of the poor to support themselves from local ecosystems while benefiting well-off farmers. The centralized nature of crop biotechnology will further this process by reducing local specificity and adaptation of agricultural practices, which increases both social dependency on external inputs to agriculture and decreases the ability of local agroecosystems to adapt to local environmental contexts. While future biotechnology may be codeveloped in local communities, as Conway proposes, it currently is not.

Furthermore, it is questionable whether technical innovation is what is needed to develop more productive agriculture. The area with the greatest current need for increased agricultural production is Africa, where the green revolution was largely a failure. It is unlikely that GM crops will eliminate the social problems that led to this failure. Conway acknowledges that a large body of social science research has demonstrated that famines are caused not by food shortages or a lack of agricultural technology, but by lack of access to food. Food access is determined by institutional characteristics such as property rights, political stability, and social security systems. Even with stable or expanding food supplies, inequality in the area of food access can lead to starvation and malnutrition.

Genetically modified crops promise to increase the productivity of poor farmers in the developing world, but so do other agricultural technologies. Rather than investing in GM crops, one could invest in organic farming, integrated pest management, water management, or crop breeding. A fair assessment of the relative merits of different agricultural practices requires a systematic understanding of these alternatives. However, there has been little systematic research on the relative ecological and economic merits of alternative agricultural systems. Agricultural research has tended to narrow its focus to single goals, such as reducing erosion or increasing crop yields, rather than regarding the management of agroecosystems as a component of regional ecosystem management.

As long as transgenic crops follow closely the pesticide paradigm, such biotechnological products will do nothing but reinforce the pesticide treadmill in agroecosystems, thus legitimizing the concerns that many scientists have expressed regarding the possible environmental risks of genetically engineered organisms. The most serious ecological risks posed by the commercial-scale use of transgenic crops are:

— The spread of transgenic crops threatens crop genetic diversity by simplifying cropping systems and promoting genetic erosion;
— The potential transfer of genes from pesticide resistant crops (HRCs) to wild or semidomesticated relatives thus creating super weeds;
— HRC volunteers become weeds in subsequent crops;
— Vector-mediated horizontal gene transfer and recombination to create new pathogenic bacteria;
— Vector recombination to generate new virulent strains of virus, especially in transgenic plants engineered for viral resistance with viral genes;
— Insect pests will quickly develop resistance to crops with Bt toxin;
— Massive use of Bt toxin in crops can unleash potential negative interactions affecting ecological processes and non-target organisms.

The above impacts of agricultural biotechnology are herein evaluated in the context of agroecological goals aimed at making agriculture more socially just, economically viable and ecologically sound. Such evaluation is timely given that worldwide, there have been over 1,500 approvals for field testing transgenic crops, despite the fact that in most countries stringent procedures are not in place to deal with environmental problems that may develop when engineered plants are released into the environment. A main concern is that international pressures to gain markets and profits is resulting in companies releasing transgenic crops too fast, without proper consideration for the long-term impacts on people or the ecosystem.

Most innovations in agricultural biotechnology are profit driven rather than need driven, therefore the thrust of the genetic engineering industry is not really to solve agricultural problems, but to create profitability. This statement is supported by the fact that at least 27 corporations have initiated herbicide-tolerant plant research, including the world's eight largest pesticide companies Bayer, Ciba-Geigy, ICI, Rhone-Poulenc, Dow/Elanco, Monsanto, Hoescht and DuPont, and virtually all seed companies, many of which have been acquired by chemical companies.

In the industrialized countries from 1986-1992, 57% of all field trials to test transgenic crops involved herbicide tolerance and 46% of applicants to the USDA for field testing were chemical companies. Crops currently targeted for genetically engineered tolerance to one or more herbicides includes: alfalfa, canola, cotton, corn, oats, petunia, potato, rice, sorghum, soybean, sugarbeet, sugar cane, sunflower, tobacco, tomato, wheat and others. It is clear that by creating crops resistant to its herbicides a company can expand markets for its patented chemicals. The market for HRCs has been estimated at more than $500 million by the year 2000 (Gresshoft 1996).

Although some testing is being conducted by universities and advanced research organizations, the research agenda of such institutions is being increasingly influenced by the private sector in ways never seen in the past. 46% of biotechnology firms support biotechnology research at universities, while 33 of the 50 states have university-industry centers for the transfer of biotechnology. The challenge for such organizations will not only be to ensure that ecologically sound aspects of biotechnology are researched and developed (N fixing, drought tolerance, etc.), but to carefully monitor and control the provision of applied non-proprietary knowledge to the private sector so as to protect that such knowledge will continue in the public domain for the benefit of all society.

Biotechnology and Agrobiodiversity

Although biotechnology has the capacity to create a greater variety of commercial plants, the trends set forth by TNCs is to create broad international markets for a single product, thus creating the conditions for genetic uniformity in rural landscapes. In addition, patent protection and intellectual property rights spoused by GATT, inhibiting farmers from re-using, sharing and storing seeds raises the prospect that few varieties will dominate the seed market. Although a certain degree of crop uniformity may have certain economic advantages, it has two ecological drawbacks. First, history has shown that a huge area planted to a single cultivar is very vulnerable to a new, matching strain of a pathogen or pest. And, second, the widespread use of a single cultivar leads to a loss of genetic diversity.

Evidence from the Green Revolution leaves no doubt that the spread of modern varieties has been an important cause of genetic erosion, as massive government campaigns encouraged farmers to adopt MVs and to abandon many local varieties. The uniformity caused by increasing areas sown to a smaller number of varieties is a source of increased risk for farmers, as the varieties may be more vulnerable to disease and pest attack and most of them perform poorly in marginal environments.

All the above effects are not ubiquitous to MVs and it is expected that, given their monogenic nature and fast acreage expansion, transgenic crops will only exacerbate such effects.

Assessing the Risks of GM Crops

Plant biotechnology offers many potential benefits to diverse groups of people. These benefits have led to the development of GM crops by private and public organizations. In response to public concern about these developments, governments around the world are creating regulations to manage the risks associated with the release of GM crops into agricultural systems. We discuss the regulation of plant introductions by comparing the introduction of GM crops to past experience with species introductions, before moving on to consider the specific properties of GM crop regulation.

The release of genetically modified organisms into the environment is frequently compared to the introduction of species into a novel environment.

The introduction of some species, especially agricultural species such as maize, wheat, and chickens, has provided enormous benefits to people. However, the costs of species introductions have been huge and largely unexpected. In the United States, approximately 50,000 nonindigenous species cause environmental damage and losses estimated at U.S. $137 billion per year. While many introduced species were not intentionally introduced, some agricultural species have escaped from cultivation. Again in the United States, 128 species of introduced crops have become serious weeds. It should be noted that there is relatively little public concern over the impacts of introduced species, which are known to be large, but substantial concern over the possible impacts of GM crops.

Introduced species offer some lessons for GM crops. Once established, introduced species are almost impossible to remove from an ecosystem, because they are continually reproducing, dispersing, and evolving. Although ecologists continue to increase their understanding of the dynamics of introduced species, they rarely manage to keep these species from spreading.

Compared with introduced species, the immediate ecological impacts of GM crops are likely to be minimal. GM crops are usually more dependent on human support than are introduced species. For example, crops typically depend on the removal of potential competitors by pesticides and mechanical disturbance. Furthermore, GM species can be engineered to be sterile or contain traits such as a reduced ability to disperse. Such technology may even help control invasive species. However, as the area and diversity of GM organisms increase, the risk that GM crops or genes may escape also increases. If introduced GM organisms possess novel traits that increase their ability to survive outside managed systems, their potential ecological impact may, in fact, be much greater than that of a comparable introduced species.

Scale and Type of Impacts

The risks associated with a GM crop depend on complex interactions among the specific genetic modification(s), the organism's natural history, and the properties of the ecosystem in which it is released. These complexities are compounded by the fact that the risks and benefits associated with a specific crop change and become more difficult to assess as the area planted increases.

As long as there are no unexpected interactions within the genome, the direct impact of specific genetic modifications of crops will probably be fairly predictable. Less direct impacts, such as the responses of surrounding ecosystems and people, are more difficult to predict. For example, the ultimate effect of a herbicide-resistant crop on agricultural practices is difficult to assess from field trials. Similarly, the complexity and cost of monitoring or conducting experiments increases with the size of the area in which the crops are being used. It becomes more difficult to predict, test, and monitor the effects of GM crops as their scale increases and their impacts become less direct.

The use of a GM crop requires a comprehensive analysis that includes a weighting of potential benefits against risks. To date, crop scientists have focused on the effects of plants in agricultural systems, while environmentalists have focused on the social and ecological consequences of the widespread use of GM crops. Consequently, these two groups are often discussing different sets of risks and benefits. This mismatch probably contributes to the rancor of the current debate.

Environmental Impacts of GM Crops

Agriculture of any type—subsistence, organic or intensive—affects the environment, so it is natural to expect that the use of new genetic techniques in agriculture will also affect the environment. The ICSU, the GM Science Review Panel and the Nuffield Council on Bioethics, among others, agree that the environmental impact of genetically transformed crops may be either positive or negative depending on how and where they are used. Genetic engineering may accelerate the damaging effects of agriculture or contribute to more sustainable agricultural practices and the conservation of natural resources, including biodiversity. The environmental concerns associated with transgenic crops are summarised below along with the current state of scientific knowledge regarding them.

Releasing transgenic crops into the environment may have direct effects including: gene transfer to wild relatives or conventional crops, weediness, trait effects on non-target species and other unintended effects. These risks are similar for transgenic and conventionally bred crops (ICSU). Although scientists differ in their views on these risks, they agree that environmental impacts need to be assessed on a case-by-case basis and recommend post-release ecological monitoring to detect any unexpected events. Transgenic crops may also entail positive or negative indirect environmental effects through changes in agricultural practices such as pesticide and herbicide use and cropping patterns.

Transgenic trees involve similar environmental concerns, although there are additional concerns because of their long life cycle. Transgenic micro-organisms used in food processing are normally used under confined conditions and are generally not considered to pose environmental risks. Some micro-organisms can be used in the environment as biological control agents or for bioremediation of environmental damage (e.g. oil spills), and their environmental effects should be assessed prior to release. Environmental concerns related to transgenic fish primarily focus on their potential to breed with and outcompete wild relatives (ICSU). Transgenic farm animals would probably be used in highly confined conditions, so they would pose little risk of environmental damage.

Gene Flow

Scientists agree that gene flow from GM crops is possible through pollen from open-pollinated varieties crossing with local crops or wild relatives. Because gene flow has happened for millennia between land races and conventionally bred crops, it is reasonable to expect that it could also happen with transgenic crops. Crops vary in their tendency to outcross, and the ability of a crop to outcross depends on the presence of sexually compatible wild relatives or crops, which varies according to location.

Scientists do not fully agree whether or not gene flow between transgenic crops and wild relatives matters, in and of itself. If a resulting transgenic/wild hybrid had some competitive advantage over the wild population it could persist in the environment and potentially disrupt the ecosystem. According to the GM Science Review Panel, hybridisation between transgenic crops and wild relatives seems "overwhelmingly likely to transfer genes that are advantageous in agricultural environments, but will not prosper in the wild ... Furthermore, no hybrid between any crop and any wild relative has ever become invasive in the wild in the UK".

Whether the otherwise benign flow of transgenes into land races or other conventional varieties would itself constitute an environmental problem is a matter of debate, because conventional crops have long interacted with land races in this way (ICSU). Research is needed to improve the assessment of the environmental consequences of gene flow, particularly in the long run, and to understand better the gene flow between the major food crops and land races in centres of diversity.

Weediness refers to the situation in which a cultivated plant or its hybrid becomes established as a weed in other fields or as an invasive species in other habitats. Scientists agree that there is only a very low risk of domesticated crops becoming weeds themselves because the traits that make them desirable as crops often make them less fit to survive and reproduce in the wild. Weeds that hybridise with herbicide-resistant crops have the potential to acquire the herbicide-tolerant trait, although this would only provide an advantage in the presence of the herbicide.

According to the GM Science Review Panel, "Detailed field experiments on several GM crops in a range of environments have demonstrated that the transgenic traits investigated—herbicide tolerance and insect resistance—do not significantly increase the fitness of the plants in semi-natural habitats". Some transgenic traits, such as pest or disease resistance, could provide a fitness advantage but there is little evidence so far that this happens or has any negative environmental consequences. More evidence is required regarding the effect of fitness-enhancing traits on invasiveness.

Management and genetic methods are being developed to minimise the possibility of gene flow. The complete isolation of crops grown on a commercial scale, either GM or non-GM, is not currently practical although gene flow can be minimised, as it currently is between oilseed rape varieties grown for food, feed or industrial oils. Management strategies include avoiding the planting of transgenic crops in their centres of biodiversity or where wild relatives are present, or using buffer zones to isolate transgenic varieties from conventional or organic varieties. Genetic engineering can be used to alter flowering periods to prevent cross-pollination or to ensure that the transgenes are not incorporated in pollen and developing sterile transgenic varieties.

Trait Effects on Non-target Species

Some transgenic traits—such as the pesticidal toxins expressed by Bt genes—may affect non-target species as well as the crop pests they are intended to control (ICSU). Scientists agree that this could happen but they disagree about how likely it is. The monarch butterfly controversy demonstrated that it is difficult to extrapolate from laboratory studies to field conditions. Field studies have shown some differences in soil microbial community structure between Bt and non-Bt crops, but these are within the normal range of variation found between cultivars of the same crop and do not provide convincing evidence that Bt crops could be damaging to soil health in the long term.

Although no significant adverse effects on non-target wildlife or soil health have so far been observed in the field, scientists disagree regarding how much evidence is needed to demonstrate that growing Bt crops is sustainable in the long term. Scientists agree that the possible impacts on non-target species should be monitored and compared with the effects of other current agricultural practices such as chemical pesticide use. They acknowledge that they need to develop better methods for field ecological studies, including better baseline data with which to compare new crops (ICSU).

Indirect Environmental Effects

Transgenic crops may have indirect environmental effects as a result of changing agricultural or environmental practices associated with the new varieties. These indirect effects may be beneficial or harmful depending on the nature of the changes involved. Scientists agree that the use of conventional agricultural pesticides and herbicides has damaged habitats for farmland birds, wild plants and insects and has seriously reduced their numbers. Transgenic crops are changing chemical and land-use patterns and farming practices, but scientists do not fully agree whether the net effect of these changes will be positive or negative for the environment (ICSU). Scientists acknowledge that more comparative analysis of new technologies and current farming practices is needed.

Pesticide use

The scientific consensus is that the use of transgenic insect-resistant Bt crops is reducing the volume and frequency of insecticide use on maize, cotton and soybean (ICSU). These results have been especially significant for cotton in Australia, China, Mexico, South Africa and the United States. The environmental benefits include less contamination of water supplies and less damage to non-target insects (ICSU). Reduced pesticide use suggests that Bt crops would be generally beneficial to in-crop biodiversity in comparison with conventional crops that receive regular, broad-spectrum pesticide applications, although these benefits would be reduced if supplemental insecticide applications were required. As a result of less chemical pesticide spraying on cotton, demonstrable health benefits for farm workers have been documented in China and South Africa.

Herbicide use

Herbicide use is changing as a result of the rapid adoption of HT crops (ICSU). There has been a marked shift away from more toxic herbicides to less toxic forms, but total herbicide use has increased. Scientists agree that HT crops are encouraging the adoption of low-till crops with resulting benefits for soil conservation (ICSU). There may be potential benefits for biodiversity if changes in herbicide use allow weeds to emerge and remain longer in farmers' fields, thereby providing habitats for farmland birds and other species, although these benefits are speculative and have not been strongly supported by field trials to date.

There is concern, however, that greater use of herbicides—even less toxic herbicides—will further erode habitats for farmland birds and other species (ICSU). The Royal Society has published the results of extensive farm-scale evaluations of the impacts of transgenic HT maize, spring oilseed rape (canola) and sugar beet on biodiversity in the United Kingdom. These studies found that the main effect of these crops compared with conventional cropping practices was on weed vegetation, with consequent effects on the herbivores, pollinators and other populations that feed on it.

These groups were negatively affected in the case of transgenic HT sugar beet, positively affected in the case of maize and showed no effect in spring oilseed rape. They conclude that commercialisation of these crops would have a range of impacts on farmland biodiversity, depending on the relative efficacy of transgenic and conventional herbicide regimes and the degree of buffering provided by surrounding fields. Scientists acknowledge that there is insufficient evidence to predict what the long-term impacts of transgenic HT crops will be on weed populations and associated in-crop biodiversity.

Pest and weed resistance

Scientists agree that extensive long-term use of Bt crops and glyphosate and gluphosinate, the herbicides associated with HT crops, can promote the development of resistant insect pests and weeds. Similar breakdowns have routinely occurred with conventional crops and pesticides and, although the protection conferred by Bt genes appears to be particularly robust, there is no reason to assume that resistant pests will not develop. Worldwide, over 120 species of weeds have developed resistance to the dominant herbicides used with HT crops, although the resistance is not necessarily associated with transgenic varieties.

Because the development of resistant pests and weeds can be expected if Bt and glyphosate and gluphosinate are overused, scientists advise that a resistance management strategy be used when transgenic crops are planted (ICSU). Scientists disagree about how effectively resistance management strategies can be employed, particularly in developing countries (ICSU). The extent and possible severity of impacts of resistant pests or weeds on the environment are subject to debate.

Abiotic stress tolerance

New transgenic crops with tolerance to various abiotic stresses are being developed that may allow farmers to cultivate soils that were previously not arable. Scientists agree that these crops may be environmentally beneficial or harmful depending on the particular crop, trait and environment (ICSU).

Assessment of Environmental Impacts

There is broad consensus that the environmental impacts of transgenic crops and other living modified organisms (e.g. transgenic seeds) should be evaluated using science-based risk assessment procedures on a case-by-case basis depending on the particular species, trait and agro-ecosystem. Scientists also agree that the environmental release of transgenic organisms should be compared with other agricultural practices and technology options.

Food safety assessment procedures are well developed and the FAO/WHO Codex Alimentarius Commission provides an international forum for developing food safety guidelines for transgenic foods. By contrast, there are no internationally agreed guidelines and standards for assessing the environmental impacts of transgenic organisms (ICSU). Scientists agree that there is a need for internationally and regionally harmonised methodologies and standards for assessing environmental impacts in different ecosystems.

According to the ICSU, regulators in different countries typically require similar types of data for environmental impact assessments, but they differ in their interpretation of these data and of what constitutes an environmental risk or harm. Scientists also differ on what the appropriate basis for comparison should be: with current agricultural systems and/or baseline ecological data (ICSU). An FAO expert consultation agreed that the impacts of agriculture on the environment were much greater than the measurable impacts of a shift from conventional to transgenic crops, so the basis of comparison is important.

Scientists also disagree about the value of small-scale laboratory and field trials and their extrapolation to large-scale effects, and it is unclear whether modelling approaches that incorporate data from geographical information systems would be useful in predicting the effects of living modified organisms (LMOs) in different ecosystems (ICSU). The scientific community recommends that more research is needed on the post-release effects of transgenic crops. There is also a need for more targeted post-release monitoring and better methodologies for monitoring.

International Environmental Agreements and Institutions

Several international agreements and institutions are relevant to the environmental aspects of certain transgenic products, among them the Convention on Biological Diversity, the Cartagena Protocol on Biosafety and the International Plant Protection Convention. The roles and provisions of these bodies are described below.

The Convention on Biological Diversity and the Cartagena Protocol on Biosafety

Most of the measures of the Convention on Biological Diversity (CBD) focus on the conservation of ecosystems; however, two aspects concerning the conservation of biological diversity are relevant for biosafety—the management of risks associated with LMOs resulting from biotechnology and the management of risks associated with alien species.

In the context of *in-situ* conservation measures, the Convention requires contracting parties "... to regulate, manage or control the risks associated with the use and release of living modified organisms resulting from biotechnology which are likely to have adverse environmental impacts that could affect the conservation and sustainable use of biological diversity ...". This provision goes beyond the general scope of the Convention in that it requires also that risks to human health are taken into account.

The Convention establishes that contracting parties have the obligation to prevent the introduction of alien species and to control or to eradicate those alien species that threaten ecosystems, habitats or species. Invasive alien species are considered as species introduced deliberately or unintentionally outside their natural habitats where they have the ability to establish themselves, invade, replace natives and take over the new environment.

The Cartagena Protocol on Biosafety was adopted by the CBD in September 2000 and came into force in September 2003. The objective of the Protocol is to protect biological diversity from the potential risks posed by safe transfer, handling and use of LMOs resulting from modern biotechnology. Risks to human health are also considered. The Protocol is applicable to all LMOs, except pharmaceuticals for humans that are addressed by other international agreements or organisations.

The Protocol sets out an Advance Informed Agreement (AIA) procedure for LMOs intended for intentional introduction into the environment that may have adverse effects on the conservation and sustainable use of biodiversity. The procedure requires, prior to the first intentional introduction into the environment of an importing party:

— notification of the party of export containing certain information;
— acknowledgement of its receipt; and
— the written consent of the party of import.

Four categories of LMO are exempted from the AIA: LMOs in transit, LMOs for contained use, LMOs identified in a decision of the Conference of Parties/ Meeting of Parties as not likely to have adverse effects on biodiversity conservation and sustainable use, and LMOs intended for direct use as food, feed or for processing.

For LMOs that may be subject to transboundary movement for direct use as food or feed, or for processing, Article 11 provides that a party that makes a final decision for domestic use, including placing on the market, must notify the Biosafety Clearing-House established under the Protocol. The notification is to contain minimum information required under Annex II. A contracting party may take an import decision under its domestic regulatory framework, provided this is consistent with the Protocol. A developing country contracting party, or a party with a transition economy that lacks a domestic regulatory framework, can declare through the Biosafety Clearing-House that its decision on the first import of an LMO for direct use as food, feed or for processing will be pursuant to a risk assessment. In both cases lack of scientific certainty because of insufficient relevant scientific information and knowledge regarding the extent of potential adverse effects shall not prevent the contracting party of import from taking a decision, as appropriate, in order to avoid or minimise potential adverse effects.

Risk assessment and risk management are requirements for both AIA and Article 11 cases. The risk assessment must be consistent with criteria enumerated in an annex. In principle, risk assessment is to be carried out by competent national decision-making authorities. The exporter may be required to undertake the assessment. The importing party may require the notifier to pay for the risk assessment.

The Protocol specifies general risk management measures and criteria. Any measures based on risk assessment should be proportionate to the risks identified. Measures to minimise the likelihood of unintentional transboundary movement of LMOs are to be taken. Affected or potentially affected states are to be notified when an occurrence may lead to an unintentional transboundary movement.

The Protocol also contains provisions on LMO handling, packaging and transportation. In particular, each contracting party is to take measures to require documentation that:

a) for LMOs intended for direct use as food or feed, or for processing, clearly identifies that they "may contain" LMOs and are "not intended for intentional introduction into the environment", and a contact point for further information;

b) for LMOs destined for contained use, clearly identifies them as LMOs and specifies any requirements for safe handling, storage, transport and use, and a contact point and consignee;

c) for LMOs intended for intentional introduction into the environment of the party of import, clearly identifies them as LMOs and specifies the identity and traits/characteristics, any requirements for safe handling, storage, transport and use, and a contact point, the name/ address of the importer/exporter and a declaration that the movement conforms to the Protocol's requirements applicable to the exporter.

Information exchange is envisaged in the Protocol through the establishment of the Biosafety Clearing-House. The Biosafety Clearing-House is intended to facilitate the exchange of information on, and experience with, LMOs and to assist parties in implementation of the Protocol. Pursuant to Article 20, paragraph 2, it shall also provide access to other international biosafety information exchange systems. Information that parties are required to provide to the Clearing-House includes existing laws, regulations and guidelines for implementation of the Protocol; information required for the AIA; any bilateral, regional and multilateral agreements within the context of the Protocol; summaries of risk assessment and final decisions.

Public participation is specifically addressed in Article 23. Contracting parties shall:

a) promote and facilitate public awareness, education and participation concerning safe transfer, handling and use of LMOs;
b) endeavour to ensure public awareness and education encompasses access to information on LMOs identified by the Protocol that may be imported;
c) consult the public in the decision-making process regarding LMOs and shall make decisions available to the public in accordance with national laws and regulations. Confidential information is to be respected in those activities.

Socio-economic considerations are allowed in decision-making. Contracting parties may account for socio-economic considerations arising from the impact of LMOs on biodiversity conservation and sustainable use, especially with regard to the value of biodiversity to indigenous and local communities. The parties are encouraged to cooperate on research and information exchange on any socio-economic impacts of LMOs. A process to address liability and redress for damage resulting from LMO transboundary movements is to be set up by the first meeting of parties to the Protocol.

IPPC and Living Modified Organisms

The purpose of the International Plant Protection Convention (IPPC) is to secure common and effective action to prevent the spread and introduction of pests of plants and plant products, and to promote measures for their control. Although the IPPC makes provision for trade in plants and plant products, it is not limited in this respect. Specifically, the scope of the IPPC extends to the protection of wild flora in addition to cultivated flora, and covers both direct and indirect damage from pests, including weeds. The IPPC plays an important role in the conservation of plant biodiversity and in the protection of natural resources. Hence, standards developed under the IPPC are also applicable to key elements of the CBD, including the prevention and mitigation of impacts of alien invasive species, and the Cartagena Protocol on Biosafety. As a consequence, the CBD, FAO and IPPC have established a close collaborative relationship. This has in particular extended to the inclusion of CBD concerns in the development of new international standards for phytosanitary measures (ISPMs).

ISPMs developed under the auspices of the IPPC provide internationally agreed guidance to countries on measures to protect plant life or health from the introduction and spread of pests or diseases. One of the most important concept standards developed under the IPPC is ISPM No. 11, Pest risk analysis for quarantine pests, adopted by the Interim Commission on Phytosanitary Measures (ICPM) at its 3rd Session in 2001. In addition, the ICPM, at its 5th Session in 2003, adopted a supplement to ISPM No. 11 to address risks to the environment in order to take into account CBD concerns, especially with regard to invasive alien species. More recently, the IPPC has drafted another supplement to ISPM No. 11 to address pest risk analysis for LMOs.

This draft standard has undergone extensive technical discussion and consultation throughout its development. At the request of the ICPM, an open-ended expert working group was convened in September 2001 and included government-nominated experts from developed and developing countries and experts representing both plant protection and environmental concerns. The purpose of the meeting was to discuss the development of this standard and the need to provide detailed guidance on conducting risk analyses to address the potential plant health effects of LMOs with particular attention to the needs of developing countries.

The working group considered that potential phytosanitary risks of LMOs that may need to be considered in a pest risk analysis include:

— Changes in adaptive characteristics that may increase the potential invasiveness including, for example: drought tolerance of plants; herbicide tolerance of plants; alterations in reproductive biology;

dispersal ability of pests; pest resistance; and pesticide resistance.

- Gene flow including, for example: transfer of herbicide resistance genes to compatible species; and the potential to overcome existing reproductive and recombination barriers.
- Potential to affect non-target organisms adversely including, for example: changes in host range of biological control agents or organisms claimed to be beneficial; and effects on other organisms such as biological control agents, beneficial organisms and soil microflora that result in a phytosanitary impact.
- Possibility of phytopathogenic properties including, for example: phytosanitary risks presented by novel traits in organisms not normally considered a phytosanitary risk; enhanced virus recombination, trans-encapsidation and synergy events related to the presence of virus sequences; and phytosanitary risks associated with nucleic acid sequences present in the insert.

Subsequently, a small working group, including CBD/Cartagena Protocol and plant protection experts, met to prepare a draft standard that would provide general guidelines on the conduct of pest risk analysis with respect to the potential phytosanitary risks identified above. In the process of drafting the standard, the working group noted several important issues with regard to the scope of the IPPC and potential phytosanitary risks of LMOs. In particular, the working group noted that whereas some types of LMO would require pest risk analyses because they could present phytosanitary risks, many other categories of LMO, e.g. those with modified characteristics such as ripening time or storage/shelf life, do not present phytosanitary risks.

Assessing the Safety of Transgenic Foods

Currently available transgenic crops and foods derived from them have been judged safe to eat and the methods used to test their safety have been deemed appropriate. These conclusions represent the consensus of the scientific evidence surveyed by the ICSU and they are consistent with the views of the World Health Organisation. These foods have been assessed for increased risks to human health by several national regulatory authorities using their national food safety procedures (ICSU). To date no verifiable untoward toxic or nutritionally deleterious effects resulting from the consumption of foods derived from genetically modified crops have been discovered anywhere in the world. Many millions of people have consumed foods derived from GM plants—mainly maize, soybean and oilseed rape—without any observed adverse effects (ICSU).

The lack of evidence of negative effects, however, does not mean that new transgenic foods are without risk. Scientists acknowledge that not enough is known about the long-term effects of transgenic foods. It will be difficult to detect long-term effects because of many confounding factors such as the underlying genetic variability in foods and problems in assessing the impacts of whole foods. Furthermore, newer, more complex genetically transformed foods may be more difficult to assess and may increase the possibility of unintended effects. New profiling or "fingerprinting" tools may be useful in testing whole foods for unintended changes in composition (ICSU).

The main food safety concerns associated with transgenic products and foods derived from them relate to the possibility of increased allergens, toxins or other harmful compounds; horizontal gene transfer particularly of antibiotic-resistant genes; and other unintended effects. Many of these concerns also apply to crop varieties developed using conventional breeding methods and grown under traditional farming practices (ICSU).

Allergens and Toxins

Gene technology—like traditional breeding - may increase or decrease levels of naturally occurring proteins, toxins or other harmful compounds in foods. Traditionally developed foods are not generally tested for these substances even though they often occur naturally and can be affected by traditional breeding. The use of genes from known allergenic sources in transformation experiments is discouraged and if a transformed product is found to pose an increased risk of allergenicity it should be discontinued. The GM foods currently on the market have been tested for increased levels of known allergens and toxins and none has been found (ICSU). Scientists agree that these standard tests should be continuously evaluated and improved and that caution should be exercised when assessing all new foods, including those derived from transgenic crops.

Antibiotic Resistance

Horizontal gene transfer and antibiotic resistance is a food safety concern because many first-generation GM crops were created using antibiotic-resistant marker genes. If these genes could be transferred from a food product into the cells of the body or to bacteria in the gastrointestinal tract this could lead to the development of antibiotic-resistant strains of bacteria, with adverse health consequences. Although scientists believe the probability of transfer is extremely low, the use of antibiotic-resistant genes has been discouraged by an FAO and WHO expert panel and other bodies. Researchers have developed methods to eliminate antibiotic-resistant markers from genetically engineered plants.

Other unintended changes in food composition can occur during genetic improvement by traditional breeding and/or gene technology. Chemical analysis is used to test GM products for changes in known nutrients and toxicants in a targeted way. Scientists acknowledge that more extensive genetic modifications involving multiple transgenes may increase the likelihood of other unintended effects and may require additional testing.

Benefits of Transgenic Foods

Scientists generally agree that genetic engineering can offer direct and indirect health benefits to consumers (ICSU). Direct benefits can come from improving the nutritional quality of foods, reducing the presence of toxic compounds and by reducing allergens in certain foods. However, there is a need to demonstrate that nutritionally significant levels of vitamins and other nutrients are genetically expressed and nutritionally available in new foods and that there are no unintended effects (ICSU). Indirect health benefits can come from reduced pesticide use, lower occurrence of mycotoxins, increased availability of affordable food and the removal of toxic compounds from soil. These direct and indirect benefits need to be better documented.

International Standards for Food Safety Analysis

At the 26th session of the Codex Alimentarius Commission, held from 30 June to 7 July 2003, landmark agreements were adopted on principles for the evaluation of food derived from modern biotechnology, and on guidelines for the conduct of food safety assessment of foods derived from recombinant-DNA plants and from foods produced using recombinant-DNA micro-organisms. A fourth document on labelling remains under discussion.

These Codex guidelines indicate that the safety assessment process for a transgenic food should be conducted through comparing it with its traditional counterpart, which is generally considered as safe because of a long history of use, focusing on the determination of similarities and differences. If any safety concern is identified, the risk associated with it should be characterised to determine its relevance to human health. This begins with the description of the host and donor organisms and the characterisation of the genetic modification. The subsequent safety assessment should consider factors such as toxicity, tendencies to provoke allergic reaction (allergenicity), effects of changed composition of key nutrients (antinutrients) and metabolites, the stability of the inserted gene and nutritional modification associated with genetic modification. If the entire assessment of these factors concludes that the GM food in question is as safe as its conventional counterpart, the food is then considered safe to eat.

Critics of this comparative approach argue that non-targeted methods that analyse the content of whole foods are needed to assess both intended and unintended effects (ICSU). Scientists generally agree that transgenic foods should be assessed on a case-by-case basis, focusing on the particular product rather than on the process by which it was created. They also agree that the safety of GM foods should be assessed before they are put on the market, because postmarket monitoring is likely to be difficult, expensive and may not yield useful data because of the complex composition of diets and genetic variability in populations (ICSU).

Risk Analysis of Modern Food Biotechnology

The risk assessment principles clarify that risk assessment includes a safety assessment designed to identify whether a hazard, nutritional or other safety concern is present and, if so, to gather information on its nature and severity. They reflect the concept of substantial equivalence whereby the safety assessment should include, but should not be substituted for, a comparison between the food derived from modern biotechnology and its conventional counterpart. The comparison should determine similarities and differences between the two. A safety assessment should (a) account for intended and unintended effects, (b) identify new or altered hazards and (c) identify changes relevant to human health in key nutrients. Safety assessment should take place on a case-by-case basis.

Risk management measures are to be proportional to the risk. These should take into account, where relevant, "other legitimate measures" according to general decisions of the Codex Commission and the Codex working principles on risk analysis. Different risk management measures can meet the same objective. Risk managers are to account for the uncertainties identified in the risk assessment and manage the uncertainties. Risk management measures could include food labelling, conditions on marketing approvals, postmarketing monitoring and development of methods to detect or identify foods derived from modern biotechnology. The tracing of the product may also be useful for the smooth operation of the risk management measure.

The risk communication principles are premised on the ideal that effective communication is essential in all phases of risk assessment and management. It is to be an interactive process stimulating advice and stakeholder participation. Processes should be transparent, fully documented and open to public scrutiny while respecting legitimate concerns for confidential commercial information. Safety assessment reports and other aspects of the decision-making process should be available to the public. Responsive consultation processes should be created.

Guideline for the Conduct of Food Safety Assessment of Foods Derived from Recombinant-DNA Plants

The Guideline for the conduct of food safety assessment of foods derived from recombinant-DNA plants was also adopted by the 26th session. The Guideline is designed to support the Principles for the risk analysis of foods derived from modern biotechnology. It describes the recommended approach for making a safety assessment of foods derived from recombinant-DNA plants where a conventional counterpart exists. A conventional counterpart is defined as "a related plant variety, its components and/or products for which there is experience of establishing safety based on a common use as food". The techniques described in the Guideline may be applied to foods derived from plants that have been altered by techniques other than modern biotechnology.

The Guideline provides an introduction and rationale for food safety assessment of recombinant-DNA plants, drawing distinctions between it and conventional toxicological risk assessment for individual compounds that rely on animal studies. The "goal of the assessment is a conclusion as to whether the new food is as safe as and no less nutritious than the conventional counterpart against which it is compared". The Guideline indicates that substantial equivalence is not a safety assessment per se.

Rather, it represents a starting point to structure food safety assessments relative to a conventional counterpart.

Substantial equivalence is used to identify similarities and differences between the new food and the conventional counterpart. The safety assessment then assesses the safety of identified differences, taking into consideration unintended effects resulting from genetic modification. Risk managers subsequently judge this and design risk management measures as appropriate.

Guideline for the Conduct of Food Safety Assessment of Foods Produced using Recombinant-DNA Micro-organisms

This Guideline is also intended to provide guidance on the safety assessment procedure of foods that are produced by using recombinant-DNA micro-organisms, based on the risk assessment framework of the above-mentioned Principles. The interesting point in the case of recombinant-DNA micro-organisms is that the comparison is recommended not only between the recombinant-DNA micro-organisms and their conventional counterparts (micro-organisms) but also between the foods produced by using them and the original foods.

Codex Text under Discussion on the Labelling of Genetically Modified Foods

In addition to the principles and guidelines above, the Draft guidelines for the labelling of foods obtained through certain techniques of genetic modification/genetic engineering are still in an early stage of discussion and many sections are bracketed, meaning the language has not yet been agreed. The guideline is proposed to apply to labelling of foods and food ingredients in three situations, when they are: (1) significantly different from conventional counterparts; (2) composed of or contain GM/GE organisms or contain protein or DNA resulting from gene technology; and (3) when they are produced from but do not contain GM/GE organisms, protein or DNA from gene technology.

According to the ICSU, scientists do not fully agree about the appropriate role of labelling. Although mandatory labelling is traditionally used to help consumers identify foods that may contain allergens or other potentially harmful substances, labels are also used to help consumers who wish to select certain foods on the basis of their mode of production, on environmental (e.g. organic), ethical (e.g. fair trade) or religious (e.g. kosher) grounds. Countries differ in the types of labelling information that are mandatory or permitted.

According to the ICSU, "labelling of foods as GM or non-GM may enable consumer choice as to the process by which the food is produced [but] it conveys no information as to the content of the foods, and whether there are any risks and/or benefits associated with particular foods." The ICSU suggests that more informative food labelling that explained the type of transformation and any resulting compositional changes could enable consumers to assess the risks and benefits of particular foods.

Weighing Risks and Benefits

Given the historical, political, and economic context of biotechnology, it is appropriate to question the completeness of current risk assessment practices. Aspects of GM crops that are relevant to the profits of farmers and agrobusiness have been well studied. However, the concerns of those who eat food, who live downstream, or who value biodiversity are only beginning to be addressed. Regulatory agencies have not been using ecologically comprehensible criteria to assess the risks associated with GM organisms. This deficiency can be explained by the political pressure that has been exerted on agencies to rapidly approve the release of genetically engineered crops. This pressure has left agencies without the physical, institutional, and conceptual framework necessary to thoroughly evaluate the risks associated with particular crops. One possible approach is to assess the risks and benefits of specific types of GM modification, for different species, in different ecological contexts.

A list of risks and benefits provides a framework that makes it easier to screen for the possible combinations of technology, crop, and ecological context that are likely to be relatively benign or hazardous. However, constructing such lists is only the first step in a risk assessment. These risks need to be quantitatively assessed for specific organisms in different contexts on a case-by-case basis. Various groups of ecologists have developed a methodology for evaluating the use of GM crops. They recommend an incremental, tiered approach to risk assessment that moves from the laboratory to greenhouse and field trials and finally to gradually increased, monitored use.

While field trials are a necessary step in evaluating GM crops, on their own they are insufficient. A more comprehensive analysis is required that includes an assessment of the relative benefits and risks of GM crops for other ecosystems and for people. Comprehensive risk assessment could allow people to reap substantial benefits from GM crops while avoiding or mitigating serious risks.

Regulating Risks

The regulatory systems designed to deal with GM crops should try to reduce the amount of risk and create the social adaptive capacity necessary to cope with the risks associated with new technologies. There are many different ways to achieve these goals. We briefly discuss three separate and complementary methods for addressing each of these challenges: biosafety protocols, a moratorium, and insurance.

Conway sensibly calls for adequate biosafety protocols in all countries. A step toward this goal was taken earlier this year when a number of countries agreed on the need for an international biosafety protocol for trade in GM organisms. However, the research and administration necessary to maintain this protocol will be expensive and must be further developed. Those who wish to produce GM crops should support the development and maintenance of a biosafety infrastructure, particularly in developing countries where it is most needed. Such support could be generated through taxes on GM crops, regulatory fees, or other mechanisms, such as a global biosafety process.

One of the current concerns is that GM crops will be widely planted before effective risk assessment procedures are in place. This has led to calls for a moratorium on the further approval of GM crops. Currently, Austria, the UK, and Germany have moratoriums, while the EU has a de facto moratorium. Such moratoriums delay the introduction of crops that could reduce the amount of ecological degradation produced by agriculture. However, moratoriums offer a number of benefits. A delay could provide the opportunity to develop institutions to effectively evaluate and monitor GM crops. It would also allow science to better assess the potential indirect impacts of existing GM crops, such as the evolution of *Bt* resistance. Furthermore, a moratorium may provide the time needed to allow a richer public debate to address how to fairly balance the risks and benefits of GM crops.

Given the uncertainty surrounding both the likelihood and degree of potential impacts of GM crops, it is sensible for society to purchase insurance against these risks. However, due to the unknown and variable nature of risks, private insurance is virtually impossible, which forces the public to play this role. Taxes on the use of GM crops could function as a type of social insurance, as long as such a tax was invested in ecological conservation and restoration, to mitigate against any disruption caused by GM crops. Insurance mechanisms should be in place before GM crops are more widely used. In addition, it would be sensible for society to invest in developing mechanisms such as research, monitoring systems, and avenues for citizen involvement to provide an early warning of any negative impacts of GM crops.

Reforming Agriculture

Reforming the scientific assessment and implementation of GM crops would be a great advance, but it will do little to solve most of the ecological problems associated with agriculture. That would require the reform of the political and economic institutions that affect agriculture.

Agricultural intensification does not necessarily have to be at the expense of ecological services. In fact, humanity can no longer afford this, because the impacts of agriculture on Earth's ecosystems are already massive However, it will be difficult to achieve productive, sustainable agriculture if society does not value ecological services. An assessment of the risks and benefits of GM crops would be greatly facilitated by efforts to value environmental services (e.g., considering the public costs of cleaning water following pesticide and fertilizer use). If these services are treated as an open-access resource, there is little incentive to conserve them. Appropriate institutions should be developed to manage the enhancement and continual provision of these services.

The economic value of ecological services is significant. A recent economic study of UK agriculture estimates that the indirect costs of agriculture, which are those paid by the nonfarming members of society, are almost as large as agriculture's net income. The presence of large costs that are not paid for by agriculture suggests that total societal benefits could be increased by moving to a form of agriculture that is less intensive and requires fewer inputs.

Public Perceptions of Safety

We join the other commentators in embracing Conway's call for a "new way of talking and reaching decisions" about development and the environment that is based upon "honesty, full disclosure, and a very uncertain shared future." However, we are concerned by his statement that the purpose of this dialogue is "to put science back at the center of a discussion of risks and benefits." Science is a necessary part of the debate over GM crops. However, a debate centered on science rather than on ethical, social, and political concerns will likely encourage further polarization and public distrust of science, which, as Krebs suggests, is a consequence of scientists' role to date in the UK debate over GM crops.

Public perceptions of safety are important aspects of social welfare. The existence of widespread public concern is an indicator of the failure of the existing system. Public concern over GM crops should not be ascribed simply to ignorance. As Krebs notes, the public acceptance of GM crops appears to decrease with scientific understanding. Skepticism over claims by regulatory agencies, corporations, and scientists that a new technology is safe is not surprising due to the many occasions when similar assurances have been proven wrong. May even suggests that concerns over GM crops in the UK can be classified into three types based on past failures: health concerns of the type associated with Bovine Spongiform Encephalopathy (BSE); ecological concerns about the transmission of unintended effects along the food chain, of which DDT provides a prime example; and concerns of the Hedgerow type about the ecological impacts of changing agricultural practices.

Risk assessment and risk management are political processes. It is difficult to determine what level of risk is acceptable for a given potential benefit, because, while the direct benefits tend to accrue to a small group of people, the risks of GM crop technology are widely distributed over the population at large. This type of asymmetry between groups is usually difficult to resolve without establishing new public institutions, because it is extremely hard to mobilize large groups of people who have experienced relatively minor losses, even when the total amount of those losses is substantial. This asymmetry further highlights the prominent roles of politics and ethics in the debate over GM crops.

Science can be used to address these concerns and questions, but it will not resolve them. Those who fund research largely determine the questions to which science is applied. Technological development is usually funded by specific interest groups, and technology is used, at least initially, for the purposes and in the ways those groups intend. GM crops have been largely developed by agrobusiness companies that seek to maximize their return on investment. These companies are part of a global market system that encourages technological and resource-intensive solutions to problems. It is not surprising that the GM crops being developed share the ecological problems of other industrial input-intensive approaches to agricultural technology. Companies have a strong incentive to conduct research and develop crops that will increase their profits, hence the focus on products that require pesticides, fertilizers, and seeds. Current research on GM organisms is dominated by the goal of producing herbicide-resistant crops. Publicly funded agricultural research should be broadened to address questions such as: How can agroecosystems be designed to improve the quality of ecological services, decrease the loss of biodiversity due to agriculture, and increase sustainability?

The new dialogue that Conway calls for should broaden the discussion of GM crops to include their political, social, and ecological contexts while searching for answers to the fundamental questions of how to maintain ecological functions, move towards sustainable agriculture, and improve people's quality of life. Science can facilitate and usefully constrain this dialogue, but the primary constraints on the development of such a dialogue will be social rather than scientific.

References

Conway, G. Genetically modified crops: risks and promise. *Conservation Ecology* 4(1): 2. 2000.

James, C. Global review of commercialized genetically modified crops. ISAAA Brief Number 8. *International Service for the Acquisition of Agri-biotech Applications,* Ithaca, New York, USA. 1998.

Regal, P. A brief history of biotechnology risk debates and policies in the United States. *Occasional paper.* Edmonds Institute, Edmonds, Washington, USA. 1999.

Rissler, J., and M. Mellon. *The ecological risks of engineered crops.* MIT Press, Cambridge, Massachusetts, USA. 1996.

8

Plant Biotechnology in Developing Countries

Biotechnology has great potential to influence and benefit agriculture, forestry and fisheries. Modern techniques of biotechnology offer the potential of moving any cloned gene from any organism into any other organism and confer much greater precision and speed in achieving results as compared to conventional techniques.

In conjunction with conventional technologies, modern biotechnology holds promise of increased and sustained productivity, efficient processing for improved product diversification and utilization, adaptation of product quality to functional requirements, and decreased reliance on agrochemicals and other external inputs. It may also promote better conservation and use of genetic resources, and environmentally friendly management of natural resources. However, the number of marketable products and their influence at the farm level still seems to be limited, but is likely to increase in the next decade.

Biotechnology also poses certain challenges. These are largely determined by how, where and when it finds application. In general, the fast-paced research, predominantly funded by private-sector investment, and use of intellectual property rights in industrialized countries are seen as evidence that the application of biotechnology will hold the key to competitiveness and comparative advantage in many fields, including agriculture and food.

Biotechnology, with its vast potential and challenges, is thus of the utmost importance to agricultural development, and hence to FAO. The science of developing biotechnological tools cannot make a distinction between developing and developed countries. However, the application of such tools in the development process requires preconditions which are easily found in many developed countries, but which hardly exist in most developing countries. The current paths of research and development have given rise to concern that the disparity in harnessing biotechnologies for agricultural and economic development may increase between industrialized and developing countries.

The preamble to the FAO Constitution requires that the Organization work towards raising levels of nutrition and the standards of living of all peoples, securing improvements in the efficiency of the production and distribution of all food and agricultural products, improving the conditions of rural populations and thus contributing towards an expanding world economy. Therefore, one of the major tasks for the Organization is to ensure that the benefits of biotechnology will be shared by people in the north and the south, in both large and small, and rich and poor countries.

In recent years, various FAO bodies and conferences such as the Tenth Session of FAO's Committee on Agriculture, 1991; the FAO Conference 1989; Sessions of the FAO Commission on Plant Genetic Resources, 1989 and 1991; FAO/CTA International Symposium on Plant Biotechnologies for Developing Countries, 1989; and the 20th FAO Regional Conference for Asia and the Pacific, 1990, have strongly recommended that FAO, as the lead UN specialized agency for food and agriculture, must respond to the potential and challenges of modern biotechnology for agriculture, forestry and fisheries.

Pattern of Use and Development of Biotechnologies

There exists a gradient of biotechnologies, depending on the degree of sophistication, complexity, stage of development and application. The lower side of the gradient comprises simpler but widely used techniques such as in vitro culture, rhizobium technology, fermentation. The upper side of the gradient includes advanced techniques involving genetic engineering. The gradient of biotechnologies may be matched with the gradient of national capabilities, economic investment and efforts to provide the possibility of choosing the appropriate techniques and approaches with the most positive impacts. The pattern of use and development of biotechnologies in developed and developing countries thus varies considerably.

The Pattern in Developed Countries

In the industrialized countries, a new pattern of biotechnological research funding has emerged. With the availability of property right protection of biotechnologies and prospects of vast markets for biotechnology products and the techniques, the bulk of the research is funded, carried out, and controlled by the private sector. For instance, since 1976 when GENENTECH, the first biotechnology company, was established, the new techniques have spawned a variety of industries that now comprise more than 400 start-up firms, more than 200 established firms that have diversified into biotechnology, and more than 200 supply firms in the United States alone. The new biotechnology industry in the United States produced pharmaceutical, diagnostic tests and agricultural products worth close to US$2 billion in 1990. A similar trend is seen in Europe and Japan. It is estimated that about 60 percent of the biotechnology research and development funding in industrialized countries is from the private sector. Thus, the private sector has been a major force in enhancing the capability of these countries in this field.

Research institutions in the public sector are now generally required to raise a substantial part of their budget from non-government sources, via contractual research, licensing agreements and royalties. This is tending to increase secrecy over research findings and to hinder free scientific communication. University professors, researchers and government institution scientists are increasingly becoming entrepreneurial and are entering private industry.

Another important trend is that large multinational corporations are purchasing smaller seed and biotechnology companies and diversifying their holdings. This allows them to develop a package sale of chemicals, seeds and equipment.

Heavy involvement of the private sector and market considerations greatly influence the topics and commodities chosen for research. Major crops, commodities and farming systems of great socio-economic importance to the developing world, but of little international market importance, do not figure in the biotechnology research agenda of industrialized countries. Furthermore, these countries are keen to reduce their production costs, increase the productivity, quality and value of their products and, thus, improve their overall competitiveness in the world market.

Patterns in Developing Countries

Biotechnology facilities are being established in most developing countries. However, the level of research, development and use of biotechnology for agriculture, forestry and fisheries in the developing countries is generally far below the level in the industrialized countries. Among developing countries, the status varies considerably. A few, such as Brazil, Mexico, India, China and The Republic of Korea, have sought to gain full scientific and technological capacity, especially in agricultural biotechnology. Others, such as Indonesia, Malaysia, the Philippines and Thailand and a few countries of Latin America, have built the capacity to apply biotechniques and develop biotechnologies useful for their agriculture and food industries. The participation of the private sector in gaining biotechnological capacity is not significant in most of these countries.

Many developing countries have inadequate funding, poor human resources and limited access to information, resulting in a relatively low-level capacity for research and technology development and exploitation, especially in the field of modern biotechnology research, which is costly and requires highly trained personnel. Most developing countries are vague as to their immediate aims in agricultural biotechnology. Few have appropriate proprietary-right protection systems or mechanisms to increase their access to protected techniques and products. Furthermore, there is negligible involvement of the private sector, which accentuates the problem of insufficient attention to biotechnology.

One of the major constraints on biotechnological development in developing countries is the poor quality and extent of higher education in frontier sciences, especially molecular biology. Besides, there are no or very weak links between universities and research institutions to reinforce each other and to utilize synergistically the scarce trained personnel resources. The university experts play a negligible role in national policy formulation, including that on biotechnology. Furthermore the universities are generally not oriented for efficiency in commercialization and are therefore not able to market biotechnology products of their own or others' inventions.

Some international and donor-country public sector institutions strive to ensure that attention is given to solving, through biotechnology, the food and agriculture problems of developing countries. Among this group are the International Agricultural Research Centres (IARCs) located or operating in developing countries. Capacity exists in several of these centres for utilization of advances in biotechnologies to solve certain problems of plant production and protection and of livestock production and health, and capacity is being built in others. For example, the International Rice Research Institute (IRRI) is incorporating modern molecular biology in its rice research activities. Its programme is linked closely to the Rockefeller Foundation Biotechnology Network. The World Bank, the International Service for National Agricultural Research (ISNAR) and the Australian Government undertook a major study on the likely impact of modern biotechnology on agriculture. This study is being followed up by the participants by an expansion in the World Bank lending for biotechnology and by ISNAR's establishing an Intermediary Biotechnology Service to provide advisory services to national agricultural research systems.

Applications, Impact and Potential

Biotechnological research and development are moving at a very fast rate. For instance, three years ago, transgenics in rice, especially in japonica rices, were considered rather difficult because of the problem of regeneration through protoplast culture. But today, a large number of transgenics both in indica and japonica types are available and are being tested at various stages. It is therefore difficult to predict the potential and impact of a given technology beyond five years or so. This section gives a brief account of the current level of technology and refers to the developments likely to occur in the next three to five years.

The potential of genetic engineering in medicine has received much attention. The potential for advances in agriculture, forestry and fishery are similarly bright. Biotechniques are already being used to create new strains of crop plants, new plant and animal diagnostic products, animal vaccines, biological pesticides and herbicides, other biological control agents, and modifications in domestic animals used for food production. In some cases, applications are being held in check by the need for still more research to ensure that there are no harmful effects (or, if there are, they are greatly outweighed by the benefits) and by the slow pace of evolution of regulations governing the release of genetically modified organisms or products for general use or acquisition.

Biotechnological research, especially genetic engineering, on problems of field and tree crops, forest species and fisheries has a relatively short history. The knowledge of biological aspects of species and ecosystems needs to be expanded in may cases before new biotechnologies can be applied effectively. Even where transfer of one or more genes can be achieved reliably, expression of the transferred gene(s) may not occur in the expected way. The search for and identification and cloning of useful genes, effective vector systems, methods of gene transfer and promoter mechanisms should be intensified.

A perspective on both the difficulties and intensity of the research effort suggests that diagnostic tools will speed solutions in the next five to ten years to certain disease problems in several crop, forest, animal and fish species, such as black sigatoka of banana and plantain, and virus/viroid diseases of coconut and rice. Improved tolerance to certain physical stresses, such as heat tolerance in potatoes, may also be achieved, but prospects for improved resistance of varieties and clones to most abiotic stresses are longer term.

Crops

Modern biotechnologies can add greater precision and speed to plant breeding. Transgenics have already been reported in 40 crop plants, including maize, rice, soybean, cotton, rapeseed, potato, sugar beet, tomato, potato and alfalfa, but the new varieties are yet to be used commercially. Near-future opportunities for commercial exploitation include vegetables and fruits (potato, tomato, cucumber, cantaloupe and squash), followed by legumes (alfalfa) and oilseed crops (oilseed rape). A good number of the transgenics are herbicide-resistant plants whose widespread use is somewhat controversial.

Wide use is currently made of tissue culture techniques for micropropagation of elite clones and for freeing planting materials of pathogens. Monoclonal antibodies are also in use as diagnostic aids in the detection and identification of viruses and viroids. Another culture and microspore culture giving rise to haploids are being used in variety improvement to facilitate and accelerate breeding. Molecular maps and markers are being widely used to identify genes of interest to accelerate conventional breeding programmes. Efficient biological nitrogen fixation systems and strains for efficient utilization of soil nutrients are being genetically engineered. Other long-term objectives are the genetic manipulations of photosynthesis pattern and the production of hybrid seed through apomixis. A very distant possibility is providing nitrogen fixation capacity to cereals.

Livestock

Among agricultural and allied fields, animal production and health have benefited the most from biotechnology, although practical use of transgenic livestock is only a future possibility. Wide use of monoclonal antibodies for efficient diagnostics, leading to safe and specific treatments of animal diseases, is a major breakthrough. Through genetic engineering, vaccines for the prevention of viral, bacterial, and parasitic animal diseases are rendered more effective and safe. Tailored vaccines exist for pig scours, chicken bursar disease and cattle tickborne diseases. Pathogen-specific vaccines are attractive goals. Other interesting possibilities are endocrine-directed vaccines to stimulate twinning in beef cattle, immunocastration, livestock growth rate increase and vaccines that compensate for various stress-induced production losses.

Advances in genetic engineering will also facilitate the production of male-only populations of screwworms, tsetse flies, ticks and various other ectoparasites for use in the sterile release technique of control or eradication. Furthermore mammalian tissue culture may replace whole animals in the 1990s for toxicity testing of certain chemicals. The culture technique can also be exploited for studying and analyzing pesticide metabolism and for herbicide pre-screening. In vitro fertilization and embryo sexing techniques have considerably increased the use of embryo transfer techniques for cattle breeding and trade. The value of the approach will be further enhanced if embryo cloning techniques can be reliably employed. Microbial and enzymatic treatment of roughages and genetic engineering of rumen bacteria both have great potential to improve animal nutrition. Growth hormones can be produced by genetically engineered microorganisms in quantities and at the low cost necessary for widespread use to speed and increase milk and lean meat production. Biotechnological tools (embryo culture, gene cloning, etc.) may also be used for conservation of genetic resources.

Forestry

Successful regeneration through micropropagation or somatic embryogenesis has been achieved for about 100 forest species although, for most, considerable development work would be required before commercial propagation could be contemplated. In breeding programmes incorporating clonal testing, inclusion of a micromultiplication phase may facilitate more rapid deployment of superior genotypes than that afforded by sequential multiplication by cuttings. This will particularly be the case when gene mappling techniques are sufficiently advanced to permit accurate identification of superior genotypes without replicated clonal testing in the field, especially within a breeding line where substantial genetic disequilibrium exists. Furthermore, gene mapping may allow eventual identification of valuable genes, including those that contribute to quantitative traits.

As regards the application of genetic engineering, many major traits of commercial importance to forestry are under polygenic control, and much remains to be learned about the operation of the genes involved before this technique can have a major impact for these characters. An earlier application of genetic engineering in some forestry operations may be the introduction of genes, some already available, known to confer insect of disease resistance. Other applications of biotechnology with obvious value for forestry, but as yet not widely supported by experimental successes, include: in vitro manipulation of the maturation state, e.g. the promotion of early flowering to reduce generation intervals, in vitro selection for traits such as disease resistance and tolerance to salinity; and the use of haploid cultures.

Fisheries

Two broad areas of biotechnology exist within the fisheries sector, namely natural products biotechnology (including mostly marine species); and aquaculture biotechnology. Commercially valuable products such as pharmaceutical compounds, pigments, oils, sterols, alginates and agarose are being extracted from micro- and macro-algae in many parts of the world. Aquatic invertebrates are currently being screened for biologically active compounds that may have, inter alia anti-carcinogenic or antiviral properties, whereas marine bacteria are currently utilized in treatments of oil spills and holding tanks in tanker ships.

Within aquaculture, induced increases in the chromosome complement (polyploidy) of commercially important species such as salmon and oysters have increased the growth and marketability of these species. Genetically engineered microbes can be used to produce fish growth hormones, which might then be used to improve feed conversion and growth rate. Synthetic reproductive hormones are produced commercially and used to regulate fecundity, breeding cycles, growth rates and sex determination in certain cultured species. In attempts to increase desirable culture qualities such as growth rate, disease resistance, temperature tolerance and marketability, transgenic fishes containing introduced genes from other species have been produced on an experimental scale. The application of advanced biotechnology to the fisheries sector is a relatively recent practice. Therefore, continued research can be expected to reveal additional commercially viable applications and products within this sector.

Processing and Product Quality and Uses

Biotechnology could have great potential for improving the quality and diversifying the uses of agricultural, forestry and fisheries products. This technology could be used for processing products by both traditional and new methods in order to:

- improve the value, safety and quality of processed and unprocessed foods, such as genetically engineered tomatoes which soften more slowly and which could be processed at lower temperatures, resulting in a higher proportion of solids to liquid in canned tomatoes;
- enhance non-food uses of farm products and thereby increase their market value;
- develop new uses for farm products, especially those often producing surpluses, including energy production from biomass and recycling of wastes and by products. An example of such an application is found in ethanol production from cassava and sugar cane;
- produce algal cells from, for instance, Spirulina or Chlorella in mass culture for proteins;
- produce enzymes on a commercial scale, especially using the microbial system, as substitutes for enzymes isolated from animal or plant sources. For example, chymosin, the enzyme used in the manufacture of cheese, traditionally taken from the cow, can now be made from specially engineered microorganisms. Mass production of amino acids from molasses and other agricultural resources is another example. These amino acids can improve the nutritional value of food and animal feed;

— produce and use new probes for rapid detection and accurate identification of specific microbial pathogens in food, especially diagnostics for staphylococcal enterotoxins, and others; and

— enhance productivity through the application of newly developed processes such as the immobilized enzyme and plant tissue culture systems which are being used to produce useful secondary metabolites. Examples are diosgenin, a female contraceptive commonly used in China and India and shikonin, a natural pigment used for lipstick in Japan.

Biotechnology is being used in a number of countries to characterize the indigenous germplasm variability for therapeutic and biologically important substances. The use of biotechnology for the removal of toxic and antinutritional factors and for improving the quality of food products is gaining momentum. These techniques include the elimination or reduction of cyanogenic compounds in cassava and the incorporation of high protein in potato and sweet potato.

Conservation and Utilization of Genetic Resources

New biotechnologies offer not only techniques to improve the conservation of genetic resources, but also new methods to identify, clone, transfer and express genes in different organisms. This has profound implications for the utilization of genetic resources, could broaden the germplasm base from which new genetic combinations can be created, and will allow scientists to pursue their breeding efforts with greater focus and speed. In vitro conservation and the use of techniques for germplasm exchange of certain plant species are already proving considerably more efficient than conventional methods. Furthermore, the generation of gene libraries provides a valuable adjunct to the conventional germplasm conservation methods. Molecular methods such as Restriction Fragment Length Polymorph (RFLP), isozyme and protein analyses are already being used for the characterization of genetic resources and the identification of useful genes. However, massive use of crop varieties reproduced through in vitro culture, which contain genetically identical copies of the parent, could provoke increasing genetic erosion.

Resistance/tolerance to Stresses

Biotechnologies have been particularly effective in and could have great potential for developing genotypes resistant/tolerant to biotic and abiotic stresses commonly affecting crop, livestock, forest and fish species, and can therefore contribute to reducing inputs of pesticides and to stabilizing agriculture in marginal lands and inhospitable habitats. Genetically engineered de novo synthesis of biopesticides by host species for their own defence, development of biocapsules containing genetically engineered biopesticides and virus coat protein-mediated resistance represent tremendous opportunities for pest management. It is anticipated that genetically engineered vegetable and cereal crops, resistant to certain insects and viruses or tolerant to herbicides, will be commercially available by 1993. Genetically engineered biopesticides are near approval for sale in some countries. The efficiency of insect-based control of pest arthropods, pathogens and weeds and the production of sterile males for insect control can be enhanced considerably through genetic engineering. Furthermore, biotechnologies are aiding conventional breeding in developing genotypes resistant/tolerant to high and low temperature regimes, drought and excess water conditions and saline and other problem soil conditions. A major concern about these solutions, however, is their sustainability if the improved attributes are based on one or only a few genes, as has often been the case.

Sustainable Production

Biotechnology may be used as a tool in the sustained production of crops, livestock, forest and fisheries by providing opportunities to:

- develop genetically broad-based varieties/breeds resistant/tolerant to biotic and abiotic stresses;
- increase efficiency in the conservation and enhanced use of natural resources, including genetic resources and systems;
- reduce use of off-farm inputs such as pesticides;
- develop populations adapted to marginal land and problem soil conditions; and
- produce genetically engineered organisms for degradation of toxic wastes and detoxification of chemical residues on produce or land.

In order to be not only environmentally friendly but also socio-economically viable and attractive, a more vigorous and focused research and development approach, especially by the public sector, is necessary. The private sector should also be encouraged to achieve the desired goal of sustainable development through the use of sound biotechniques.

Biotechnology and Intellectual Property Protection

The new biotechnologies are to be viewed as among the latest of the tools or means which advancing science has provided for the development of agriculture, forestry and fisheries. While they are already making significant contributions to pharmaceuticals, diagnostics and human health, the potential benefits are not being tapped adequately in agriculture and allied fields. In the developing countries, the overall use of modern biotechnology is rather limited.

The full value of products and technologies will only be realized when the necessary research and development infrastructures, guidelines and regulations, financing and public policies are in place. A technology will only succeed in an environment whose social and economic policies are prepared to support it. Furthermore, the products and technologies should be accessible to farmers, forests, fisherfolk and other bona fide users in both developed and developing countries. Moreover, the development gains accruing through the use of biotechnologies should be consistent with sustaining of the environment.

The recent trend of development and application of modern biotechnologies has raised certain socio-economic, institutional, environmental and political issues. Foremost among them are intellectual property protection, inadequate research and institutional support, biosafety and other environmental aspects, substitution of developing countries' exports, and social equity.

The management of intellectual property has emerged as a major policy issue affecting the development, application and diffusion of biotechnologies. Intellectual Property Right (IPR) systems have been instituted in many, mostly developed, countries, to stipulate innovations which may contribute to improved productivity. These promote investment in research and development by, and secure rewards for, the private sector. They also provide an incentive to disclose details of the innovations.

Notwithstanding the positive effects of adoption of IPR in intensive production systems with high capital input, there are certain aspects of IPR which may hinder developing countries from reaping the full potential benefits of biotechnology. Transnational corporations have been increasing their control on biotechnology research and its fruits through the acquisition of companies, patents and licences. To improve that competitive edge in international markets, governments of industrialized countries are at the forefront of international efforts to reinforce the IPR systems. The developing countries are under pressure to adopt the IPR systems proposed by the industrialized countries in order to open markets for the protected products. In some countries which have adopted "foreign" IPR systems, the vast majority of patents are held by foreign companies and are seldom used to promote local research and production systems. On the other hand, patent incentives may be useful for the more advanced developing nations and for the application of privately derived technologies.

Most developing countries, especially the least developed ones, have more need to absorb and diffuse as widely as possible the new technologies for their development. They try to avoid the overpricing and monopolization that could occur through imposition of strict IPR systems. Some even consider that the current international patent regime works to their disadvantage and that they receive nothing in return for protecting inventions produced in the developed countries.

The patenting of life-forms is yet another sensitive issue and has complex ethical and legal implications, besides economic considerations, as discussed in a later section.

Thus, the role of IPR and the private sector will vary according to socio-economic settings and government policies. A system suitable for industrialized "seed-rich" countries may not necessarily be the most apt for "gene-rich" developing countries which are mostly comprised of resource-poor farmers and consumers. Each country should, therefore, develop its own policy on IPR and its support to the private sector according to its own national development needs and capabilities. In fact, several developing countries are reviewing their approaches to IPR systems and are trying to evolve and adapt patterns befitting their aspirations and opportunities.

The Plant Breeders' Rights (PBR) system evolved by the International Union for the Protection of New Varieties of Plants (UPOV) Convention, maintained the fundamental principle of unrestricted access to genetic resources. It also had provisions for "Breeders' Exemption" (breeders can use a protected variety for creating and commercializing a new variety) and "Farmers' Privilege" (farmers are permitted to multiply propagation material of a protected variety to be used for further crop growing on their own premises). However, the recently revised UPOV Convention (April 1991) has eliminated both "Exemption" and "Privilege". The extension of the industrial patent system to plant genetic resources, as stipulated in the new UPOV Convention, will interfere with FAO's proposed balance between PBR and "Farmers' Rights". The concept of "Farmers' Rights" was approved by all Member Nations of FAO in 1989 and is defined as "the rights arising from the past, present and future contributions of farmers in conserving, improving and making available plant genetic resources, particularly those in the centres of origin/diversity". If the patent system is applied universally to living matter, including plants and animals, and their genetic resources, then the principle of unrestricted access, as stipulated in the International Undertaking on Plant Genetic Resources, will be severely eroded.

Inadequate Research and Institutional Support

As stated previously, research and technology development in biotechnology is expensive and requires highly trained personnel and comprehensive communication systems. The lack of expertise and research and development infrastructures, and the absence of appropriate policies, strategies, and priority-setting mechanisms, render most developing countries incapable of benefiting from the full potential of biotechnology. Other factors, such as the timely availability of credit, inputs and extension services, marketing and appropriate prices, are among the many economic conditions for the adoption of biotechnology, as for any new technology. A number of social and cultural conditions will also be relevant. Deficiencies in the involvement of the private sector in its links with the public sector, in policy-makers' awareness of technical possibilities, and in assessing the need for a biotechnological approach, the costs involved and the chances of success, aggravate the problems of the non-judicious use of biotechnology in most developing countries.

Advances in knowledge and applications of biotechnologies will continue and will probably accelerate in the most highly industrialized countries, who already have the lead in biotechnology development. Developing countries are therefore challenged to keep aware of the advances and to make their own appropriate contributions to developing technology. Although some laboratories in developing countries could make front-line research contributions, their priority needs would be:

- to know what is becoming available;
- to evaluate available technologies for technical suitability for removing or ameliorating existing problems or inefficiencies, or to seize worthwhile opportunities;

— to make cost/benefit assessments or informed judgements on the consequences of introducing particular biotechnology in social, economic, environmental and institutional terms and in regard to sustainability; and

— to acquire, to adapt and to bring to application in the fields of agriculture, forestry and fisheries those biotechnologies which will improve the welfare of the country and its people.

The transition from research and development of technologies to commercial application is rather bumpy and inefficient in most developing countries. These countries should devote greater attention to applied research and technology transfer and should develop suitable personnel, infrastructures and institutions to facilitate technology application. For a workable transition to be made from the scientific effort to the market, a wide variety of capabilities and institutions have to be in place, and due attention should be paid to the development of biotechnological and engineering skills, market analysis and scale-up issues.

There are great differences in biotechnological capabilities among developing countries as well as between developed and developing countries. These differences will affect measures that the developing countries can take on their own and the extent and type of assistance they will find useful for the further development of biotechnology. Policies and measures that promote intercountry collaboration, based on the respective strengths and weaknesses of the cooperating countries, should be established and promoted in the spirit of technical cooperation among developing countries and also between developed and developing countries.

Biosafety and Other Environmental Aspects

Another major issue affecting the role and application of biotechnology relates to the safety of Genetically Modified Organisms (GMOs) and the regulatory measures for research into and field testing and commercialization of GMOs. Fears have been expressed that uncontrolled release of GMOs might cause changes in ecological or genetic equilibria, with unforeseeable and perhaps deleterious consequences. This issue has been considered by several national and international bodies, including the US National Academy of Sciences, the US Congress Office of Technology Assessment, the European Commission, the Organization for Economic Cooperation and Development (OECD), the Australian Recombinant DNA Monitoring

Committee, the UK Royal Commission on Environment Pollution and The United Nations Environment Programme (UNEP) and they generally believe that the technologies are safe. Studies by these organizations and others, and field tests carried out in the United States have concluded that careful design of transgenic organisms, along with proper planning and regulation of environmental releases, are sufficient to endure that GMOs will pose little or no risk to the environment.

It is increasingly appreciated that while biotechnology provides powerful new tools, they are generally used to generate products that fill essentially the same sorts of roles as those produced by more traditional methods. However, the potential power of biotechnology in creating diverse products is far-reaching and calls for greater preparedness to assess their acceptability from the point of view of biosafety. The efficacy and application of standards and regulatory measures for food and agricultural products, including food safety assessments and quarantine laws, should be examined in the context of new biotechnological products If necessary, suitable changes or strengthening measures should be brought about to render them effective both for encouraging the use of new products and for ensuring human welfare and health and environmental safety. A tiered system of responsibility for biosafety of products produced through biotechnology, at institutional and national levels, should be established.

It should also be appreciated that risks associated with the release of genetically engineered organisms cannot be reduced to zero. Biosafety aspects of genetically engineered biopesticides and other biochemicals should be assessed on the same lines as those of synthetic chemicals. Relatively greater risks may be involved if pathogenicity or weediness are implicated in release of GMOs. The risk of transferring herbicide resistance into weeds through the widespread cultivation of allied herbicide resistant transgenic crop plants should be analysed critically. The implications of releases should therefore be systematically investigated before the release is authorized, and steps should be taken to minimize potential problems.

The safety implications, if any, of applications of new biotechnologies to food production and processing should also be examined. Any safety assessment strategy should be based on considerations of the molecular, biological and chemical characteristics of the material at issue, and toxicological studies in animals. Facilities for adopting such an approach in most developing countries are inadequate. Action at the international level will be necessary to provide timely expert advice and technical assistance in this matter.

The biosafety aspects are primarily a matter for national decisions. The national governments and regional programmes should act to safeguard their own ecosystems, genetic resources and the health and well-being of their citizens from any possible risks associated with the release of GMOs and commercialization of genetically engineered food and other products. For this purpose, national governments should establish appropriate policies, laws, regulations and enforcement mechanisms for the control of potentially problematic introductions, either for testing, for export and import, or for releases on a commercial scale. Laws, regulations and guidelines on research, release, containment and monitoring of GMOs and on food safety assessment and their effective implementation are available and feasible in most industrialized countries. Generally, such measures do not exist in most developing countries because of insufficient scientific expertise and resources to assess the risks and implications adequately. However, national and international efforts are in progress to bridge this gap. International effort is further needed to develop, harmonize and implement procedures and standards based on sound and comprehensive scientific assessment, so that all countries may follow internationally agreed biosafety procedures and national decision-making processes.

Substitution of Developing Countries' Exports

One of the potential negative impacts of biotechnology for developing countries is the speeding up of the process within industrialized countries of the substitution of products or high-value components of specific products originally derived from the produce of developing countries, thus depressing the limited opportunity for exports by the latter. Several significant agricultural exports of developing countries are already threatened. For instance, a number of companies in developed countries are now using biotechnology to produce a natural vanilla flavour in the laboratory - a process which could eliminate the need for traditional cultivation of the vanilla bean. This could result in the loss of over US$50 million in annual export earnings from Madagascar, jeopardizing the livelihood of some 70 000 small farmers.

Tissue culture production of certain components could displace their field production. For instance, biotechnology laboratories in Europe, the

United States and Japan are standardizing techniques to substitute cocoa butter with cheaper vegetable oil, which could adversely affect the cocoa economy in several developing countries, especially in Africa. Other high-value, low-volume products such as pharmaceuticals, fragrances, flavourings and spices are also targets of biotechnology research. Efforts are being made to reduce the cost of production of these products under in vitro conditions in order to render the approach more competitive. There are no easy solutions to these problems. Certainly economic forces will dictate. One possible solution may be political, e.g. for production facilities to be located in nations whose economy will be affected rather than in the industrialized country that develops the process. Another solution is to reduce the cost of production in vivo by amplifying the genes responsible for production or amplifying the gene product through the incorporation of promoters in the plant genome.

Biotechnological advances in agricultural product processing have also led to the separation of plants from their specific characteristics, resulting in the substitution of one product for another. An instance of this is the increasing competition between sugar and starch producers. The production of alternative sweeteners has already adversely affected the sugar industries in major sugar-producing countries, threatening the livelihood of an estimated eight to ten million people in the developing countries by the loss of traditional sugar markets and a drop in world sugar prices. The development of High Fructose Corn Syrup (HFCS) from maize in the United States resulted in a drop in sugar exports from the Philippines to the United States from US$624 million in 1980 in US$246 million in 1984. Moreover, 500 000 people were rendered jobless because of this shift.

Social Equity

Biotechnology, like any new technology, can be neutral, pro-rich or pro-poor, depending on the stage of its development and management, area of application and the socio-economic climate in which it operates. As regards research and generation of new techniques, biotechnological techniques are certainly high-cost and require highly trained personnel setting generally obtained in industrialized countries. Much of the research takes place in the private sector, which means that the marketability of the product and potential return on investment are crucial factors in deciding what research to undertake. Most of the new technologies, processes and products are, therefore, generally expected to be first available in industrialized countries and first applied to the commodities and priorities favoured by those countries. Furthermore, research and development goals of the private sector may diverge from those of the public sector. The gap should be narrowed, keeping in mind the interest of the people, especially the poorer sector who constitute the majority in developing countries.

The adoption of biotechnology-derived products/techniques intended for common use should ordinarily be scale-neutral. However, the time lag in availability and adoption of new technologies between developed and developing countries is likely to reduce the competitiveness of agriculture in the poorer countries, and of the poorer sectors within a country, at least in the short term, especially when more and more production surpluses will be competing in fewer markets. The share of industrialized countries in the world export of food products increased from about 45 percent in the early 1960s to about 68 percent in the early 1980s.

Another anticipated consequence of the application of biotechnology is an acceleration in the trend towards further industrialization of agriculture, for which many of the smaller and less developed countries are not sufficiently prepared. Even though biotechnologically modified varieties, breeds and microorganisms may be used with equal success in small- and large-scale agriculture, economies of scale in marketing and processing and ability to take risks and to invest favour adoption first and foremost by larger producers.

Biotechnology can positively be pro-poor. It may help to create new markets, both through the breeding of new industrial, medicinal and aromatic crops, and through changes in downstream processing. Given their richness in indigenous biodiversity, several developing countries, such as Brazil, China and India, which have biotechnological capabilities, should use the new technology for production of new high-value pharmaceutical and industrial products based on their local flora. The congenial agroecological settings and the availability of relatively cheap labour should be conducive to large-scale production of new high-value crops, especially medicinal and aromatic, enabling such countries to maintain comparative advantage in these commodities. The use of biotechnological techniques for the development of biofertilizers, biological management of pests, detection of pathogens and their biocontrol, etc., which are scale-neutral and labour-intensive, will be particularly suitable for resource-poor farmers. However, their transfer will require high-quality management, which requires complementary changes in training and extension activities. These technologies will also address non-tariff trade barriers arising because of pesticide residues in or pest infestation of traded food and other products, and would thus increase the access to particular export markets.

Ethical and Legal Aspects

Notwithstanding the high potentials of biotechnology for development, the genetic manipulation of crops and livestock using genes from unrelated organisms and the possible implications for biosafety and human health have raised ethical issues. In a widespread debate in both the industrialized and developing countries, many consider that patenting higher life forms and the genetic materials they contain is unacceptable for ethical, social and economic reasons. Most developing countries do not allow the patenting of plants, animals or their genetic component generally because of their importance in the food supply. Others, however, see the use of patents as one of the necessary mechanisms to stimulate technology development.

Differences in perspectives on the usefulness and exploitation of biotechnology emanate from the level of agricultural and economic development, the level of research and technology capability, the form and mechanics of transfer of technology, and the availability of appropriate regulations and the mode of their implementation. Several technical and legal problems related to biosafety and patenting of living organisms and their genetic materials remain unresolved. The definition of protected subject-matter as it applies to biological material is still evolving and is far from fixed, and in many countries a policy debate on this matter is under way. Specific legal provisions in the area of IPR for biological content are currently under consideration in various international fore such as the World Intellectual Property Organization (WIPO), General Agreement of Tariffs and Trade (GATT) and OECD. Specialized UN agencies and technical bodies, such as FAO and World Health Organization (WHO), are closely associated with, and involved in, such negotiations/debates to facilitate the formulation of socio-economically scientifically and ethically balanced decisions.

The system which is now emerging in some industrialized countries is one which will grant strict intellectual property protection to a wide array of biotechnological products. Great pressure is being and will continue to be exerted by these countries in a number of fore in order to assure that such protection is observed worldwide, although, at the Uruguay Round of the GATT negotiations, industrialized countries formally agreed that they did not expect developing countries "to make contributions which are inconsistent with their individual development, financial and trade needs". The system proposed by industrialized countries is designed to serve individual inventors in a formal research setting and to "protect" the "inventions" originating from their own economic systems. These laws do not take into account informal research and innovations or the indigenous knowledge and products of differing cultures, which have provided and will continue to provide invaluable information and materials for further innovations worldwide.

Member countries should carefully assess the implications of different rules regulating the development and use of biotechnology. Appropriate and workable legal frameworks on biotechnology-related matters should be formulated to ensure the balanced exploitation of new techniques and products. FAO, based on its experience on issues such as the safe and efficient use of pesticides and harmonization of quarantine principles and procedures, would be one of the intergovernmental bodies, such as WIPO and WHO, to assist Member Nations in formulating the necessary legal guidelines.

FAO's Policies and Strategies

FAO considers that modern biotechnology is already making important contributions to and poses great potential and challenges for agriculture, forestry and fisheries development. The Organization recognizes that biotechnologies are a new group of powerful tools for research and ultimately for accelerating development and not an end in themselves. It perceives that modern biotechnologies should be used as adjuncts and not as substitutes to conventional technologies in solving problems, and that their application should be need- rather than technology-driven.

Successful development and application of biotechnology are possible only when a broad research and knowledge base in the biology, variation, breeding, agronomy, physiology, pathology, biochemistry and genetics of the manipulated organism exists. Benefits offered by the new technologies cannot be fulfilled without a continued commitment to basic research. Biotechnological programmes must be fully integrated into a research background and cannot be taken out of context if they are to succeed. With this background, FAO's strategy is to keep biotechnologies in a balanced perspective by undertaking activities within the framework of existing national research agendas and priorities through consultations, monitoring, and programme initiatives, rather than to support development of new independent programmes and structures around a set of technologies which, in fact, are tools to be used by diverse disciplines and programme areas.

Each country has a responsibility to formulate its own policies, priorities, strategies and programmes for harnessing biotechnology, and to weigh expected benefits, not only against possible negative effects but also against the risk of not exploiting the technology. Commensurate with these responsibilities, the countries must have the necessary infrastructures, financial support and expertise. A majority of the developing countries lack these prerequisites and will need assistance to strengthen their overall capabilities in biotechnological research and development in order to meet the potentials and challenges of the new technologies. On request, FAO can provide technical inputs to assist in planning, programming, priority setting and strategy formulation.

In line with its mandate and the three major areas of its programme, namely, providing information, providing a forum for international debate for issues related to food and agriculture, and rendering technical assistance to its Member Nations, FAO seeks, within its means and resources, to realize fully the positive impacts of biotechnologies and to minimize, if not completely eliminate, the negative effects. In this resolve, FAO's strategy is to concentrate on activities such as providing information, monitoring and advice, facilitating access to the new technologies, providing a forum for the review of trends, developing appropriate guidelines and codes to facilitate the environmentally sound and quotable harnessing of modern biotechnologies, assisting developing countries to identify biotechnology needs and priorities and to assess socio-economic impacts, and strengthening the overall biotechnological capabilities of the developing countries.

Monitoring and Appraisal

FAO, in cooperation with concerned national and international institutions, through country-level and intercountry studies and consultations and seminars, will seek to assess the realistic potential of the new biotechnologies (short-term benefits must not be overstated) as well as their limitations for agricultural, forestry and fisheries production, with special reference to developing countries. The objectives of positive benefits include increased yields, stabilized production, improved ecological sustainability, better food quality and reduced losses. FAO, also through the work of the Commission on Plant Genetic Resources, shall monitor and define potential and real adverse effects of biotechnology and develop early warning systems to ward off possible negative effects. Longer-term activities would include assessment of the propensity of widespread uses of biotechnologies to modify established patterns of comparative advantages in food, agriculture, forestry and fisheries production and trade, and to displace traditional commodities, and the implications thereof. FAO will also monitor legal, socio-economic, and other policy aspects of biotechnological advances and alert Member Nations of the implications.

Information

It will be incumbent upon FAO to collect, consolidate, analyze and disseminate information on biotechnology in relation to food, agriculture, forestry and fisheries so that the Organization can assist Member Nations in evolving appropriate policies, strategies and management systems for rational exploitation of biotechnologies. Use will be made of AGRIS and CARIS (the FAO's information databases) and Member Nations will be assisted and encouraged to participate in information gathering and exchange. FAO will also assist in the development of international centres for the genomic databases of economically important organisms, and this information will be freely disseminated.

The participants in AGRIS and CARIS are conscious of the importance of biotechnology and already, at a general consultation in May 1990, have urged each other to make greater efforts to ensure complete coverage of the available information. However, because many disciplines are involved, biotechnology information tends to be widely dispersed in the scientific literature. Therefore, it is envisaged that AGRIS inputters could use new codes to flag relevant material and thus facilitate its extraction and presentation. Specialized products could then be issued in cooperation with other units of FAO for distribution to the global and regional networks. Such products could include printed bibliographies, similar to Animal biotechnology, specialized databases on diskettes or CD-ROM for use with personal computers, and alerting services directed to individuals and research groups.AGRIS and CARIS are also exploring the possibilities of an electronic mail/bulletin board service which would enhance the speed at which critical information is delivered and provide an electronic conferencing capability for scientists in the various networks.

Increasing Access to New Technologies and Products

In order to increase access of developing countries to biotechnological techniques, processes and products, including the protected ones, FAO will provide a natural forum to review and discuss the issue of free availability and other biotechnology access issues. Various FAO bodies such as the Committee on Agriculture, the Council, the Commission on Plant Genetic Resources and the Conference, will seek formulation of agreed norms, codes and procedures for the unhindered flow of information, technology and genetically engineered materials to all bona fide users, with due provisions for biosafety and environmental soundness. FAO would serve as an intermediary between the technology developers and the users to secure or acquire for developing countries the potential benefits arising from biotechnology research and its application. For this, FAO will act as an "honest broker" to match the needs of the developing countries and appropriate the proprietary technologies they require. The Organization will mobilize funding support from donors and development banks to implement the brokered proposals.

For some time, FAO has been debating the issue of free availability of genetic resources, including genetically engineered materials, to all bona fide users, particularly since 1983 when the Organization evolved a unique global system on plant genetic resources consisting of the International Undertaking, the Commission and the International Fund. In recent meetings of the FAO Commission on Plant Genetic Resources, and at the FAO Conference, the principle that plant genetic resources are a common heritage of the human race has been further clarified: it has been stressed that "free access" does not mean free of charge and it has been pointed out that the principle of a common heritage is not incompatible with national sovereignty. The discussions surrounding the recognition of Plant Breeders' Rights and Farmers' Rights, and the establishment of the International Fund for Plant Genetic Resources, have turned towards the need to establish a mechanism, or mechanisms, to reward breeders and to compensate farmers throughout the world - especially in developing countries - for having developed and preserved, over generations, the plant genetic resources that are now being utilized, and for making those resources available to today's breeders and scientists.

In FAO's view, an IPR system should be flexible enough to match the development goals, policies and socio-economic priorities of individual countries, and at the same time should be able to promote innovation and application of new technologies. Each country needs to weight the benefits and costs of IPRs in biotechnology, and frame its policies accordingly. If the national capability in genetic improvement and biotechnology research is negligible, the country will benefit little from establishing an IPR system generally used in industrialized countries. On the other hand, where the national capability is fairly developed and agriculture is market-oriented, a moderate system of the Plant Breeders' Rights (PBR), which would stimulate private sector investment, may prove appropriate. Developing countries should opt for a PBR system with provision for farmer plantback, especially where farmers cannot afford to buy seed each year or are not reached by a seed distribution system. Finally, member countries should ensure that no IPR system be established that would restrict access to and use of genetic resources, inhibit promotion of research and technology development and thereby adversely affect their vital socio-economic fabric. The important role that farmers had, and continue to have, in the development and maintenance of germplasm should be duly recognized. FAO is involved in preparing a code of conduct for biotechnology as it relates to conservation, exchange and use of genetic resources.

Strengthening National Capabilities

Based on the needs and opportunities of individual developing countries, FAO will seek to strengthen national capabilities and to promote self-reliance for the generation and transfer of biotechnological techniques by supporting personnel development; enriching information collection, consolidation and dissemination; augmenting required research facilities; and assisting in the formulation of appropriate national policies and priorities.

— *Research and technology development*. FAO has assisted several relatively well-positioned developing countries to assess their biotechnology requirements and prepare the way through institutional and human resources development for undertaking research and technology development with special emphasis on the integration of biotechnology. Assistance has been focused on biotechnology for crop improvement, or the production of vaccines and use of diagnostic, with involvement of the Joint FAO/International Atomic Engergy Agency (IAEA) Division (AGE), in livestock health and reproduction programmes. FAO's in-house capacity will be strengthened in order to meet the demands on the Organization in the field of biotechnology effectively.

— *Personnel.* The present and future use of biotechnology depends, to a considerable extent, upon the availability of trained researchers, bioprocessing engineers and technicians. One of the major factors limiting the use of biotechnology in developing countries is the acute shortage of appropriately trained staff. FAO, through both its regular and the Field Programmes, should support developing countries in establishing functional applied biotechnology laboratories and in related training activities. The Joint FAD/IAEA Division (AGE) laboratory has a strong research-training component on biotechnologies and has been actively involved in staff training and basic research in this field. Fortunately, the techniques and methods have become more accessible through technological innovations such as the polymerase chain reaction and automated sequence analysis. Nonetheless, FAO should continue giving high priority to training scientists in developing countries.

— *Policy.* In view of the fast pace of biotechnological development and the widening gap in the capabilities of developed and developing countries, the effort to narrow this gap will be further intensified by increasing the policy-formulation, priority-setting and programme-implementation capabilities of developing countries. In particular, FAO will reinforce its assistance to developing countries in analyzing the necessary and sufficient social and economic policies for the development, adaptation and implementation of appropriate technology.

Establishment of Reference Laboratories

FAO strongly advocates observation of all internationally established norms and requirements for bioengineered vaccines and other bioproducts based on the application of microbiological factors. The Organization will continue and further strengthen its programme on the establishment of collaborating laboratories that provide invaluable services, not only as reference for quality testing but also for training. For instance, arrangements are being made to test recombinant vaccines under the FAO umbrella. In addition to strengthening national institutes and supporting ongoing efforts in developing countries, FAO has established a Collaborating Centre for Biotechnology Transfer at the University of California, Davis. A counterpart Molecular Virology Unit for Biotechnology Transfer will be organized at the Pan-African Vaccine Centre in Addis Ababa, Ethiopia. The FAD/IAEA laboratory at Seibersdorf, Austria, will be strengthened to serve as an important reference centre for new biotechniques, genetically engineered organisms and resulting products.

Promotion of Intercountry Cooperation and Networks

FAO will promote intercountry and interlaboratory cooperative research through the establishment of cooperative networks. Examples of such networks include the Asian Network for Biotechnology in Animal Production and Health; the Computer Assisted Analysis of Nucleic Acids and Protein Sequences (CANAPS) in Latin America and the Caribbean, and the Latin American Technical Cooperation Network on Plant Biotechnology. An Asian Network on Plant Biotechnology, and a Latin American Network on Animal Biotechnology, are currently being established. FAO is also sponsoring regional networks on the preparation and distribution of diagnostic probes/monoclonal antibodies for the rapid diagnosis of livestock and poultry diseases.

Based on their capabilities and willingness, certain laboratories in selected National Agricultural Research Services (NARS) will be strengthened to take the leadership role in specific areas. Results and products from these centres will be shared freely among all cooperators in the networks. FAO will work closely with selected biotechnological research and development centres both in developed and developing countries, and will seek their assistance in programme and project formulation, appraisal and evaluation.

Several expert consultations organized by FAO to prepare the ground for cooperation networks in animal production and health have recommended approaches and steps for FAO in this field. Regional expert consultations were held in the Asia-Pacific Region in 1990 and 1991 in order to establish a cooperation network on plant biotechnology. Through the 1991/92 André Meyer Fellowship on Forest Biotechnologies, the potential role of biotechnology, within well-founded tree improvement and breeding strategies, and in forestry in general, is being analyzed, with special reference to potential in developing countries.

FAO will stimulate development-related research in industrialized countries and help establish links with institutes in the developing countries. It will strive to influence at least the public sector institutions in developed countries to include in their biotechnology research agenda commodities and techniques of importance to developing countries. The Organization will encourage the strengthening of biotechnology capabilities at the IARCs, and work in concert with them in human resources development programmes and in cooperative research activities to strengthen the biotechnology capabilities of developing countries. FAO will keep abreast of opportunities for solving recalcitrant problems through the application of biotechnology, and will organize workshops/seminars to discuss methods and strategies for approaching such problems through cooperative research programmes.

Promotion of Biosafety and Environmentally Friendly Use

FAO will continue to cooperate with United Nations Industries Development Organization (UNIDO), UNEP and WHO in the establishment of codes and guidelines for biotechnology-related environmental and health risk assessment. The UNIDO/UNEP/WHO/FAO Working Group on Biosafety brought out a Voluntary Code of Conduct for the Release of Organisms into the Environment in 1991. The code sets out general principles, a framework and guidelines to be adopted at national, regional and international levels to facilitate the safe application of biotechnology. FAO will lead the establishment of "prior informed consent" and seek to further the use of relevant data to assist developing countries in elaborating pertinent policies and regulations.

FAO and WHO will jointly monitor the food safety aspects of foods prepared through biotechnology applications in accordance with established principles for the evaluation of food safety in general. While the 19th Session of the Codex Alimentarius Commission in 1991 had endorsed the opinion of the Joint FAD/WHO consultation on this subject in Geneva in 1990, that foods derived from "modern" biotechnologies were inherently not less safe than those derived from traditional biotechnologies, the issue of safety along with nutritional concerns had to be considered. The Consultation had recommended that FAO and WHO, in cooperation with other international organizations, should take the initiative in ensuring a harmonized approach on the part of the national governments to the safety assessment of foods produced by biotechnology. Based on scientific and technical advice by joint FAD/WHO committees and consultations, the various Codex Committees will discuss the issues concerned and help to reach international consensus on particular novel foods. For instance, the Codex Committee on Food Labelling is to provide guidance on how the fact that a food is derived from "modern" biotechnologies can be made known to consumers.

The Fourth Session of the FAO Commission on Plant Genetic Resources, in April 1991, resolved that FAO, in collaboration with other concerned organizations, should develop a Code that includes and promotes basic biosafety standards for the contained use and deliberate release of GMOs, and for their importation and exportation. The Code will include elements to minimize the negative potential of GMOs on genetic diversity. Work in this direction and on the lines suggested is already in progress.

Promotion of Pro-poor Features

Within the framework of overall strategies, FAO will seek to promote biotechnological research and development of commodities, often referred to as "orphan commodities", which are generally not researched by industrialized countries but which are of vital importance to most developing countries. It will strengthen national capabilities for biotechnological solutions of problems related to increased and sustained agricultural production under rainfed, saline and other marginal conditions, which are largely inhabited by resource-poor farmers and which are characterized by low and unstable productivity. FAO will also seek to strengthen national capabilities for exploiting in vitro culture techniques, biological nitrogen fixation, genetic resistance to pests and diseases, low-cost production, and increased employment opportunities. Furthermore, through expert consultations, information exchange and technical assistance, the Organization will promote the use of biotechnology for diversified and new uses of rich indigenous variability, especially those of industrial, medicinal and aromatic plants and small animals, and for developing new farming systems. These policy elements, if linked with appropriate action plans, should go a long way towards promoting equity coupled with sustained and environmentally sound economic development.

Links with other UN agencies and with NGOs

FAO has been collaborating with several UN and non-UN agencies/ systems in a number of activities related to biotechnology. For instance, it takes part in the UNEP-initiated effort to establish a Convention on Biodiversity and in the Biotechnology Working Group formed in that effort. FAO was a member of the Working Group on Biotechnology which was created in preparation for the 1992 UN Conference on Environment and Development (UNCED), and will continue to work with other UN Agencies in implementing the Action Plan on Biotechnology contained in UNCED Agenda 21. The Organization has been working with UNIDO, UNEP and WHO in developing a Code of Conduct for Biosafety, and in developing food safety standards. It will continue to work closely with WIPO and GATT in evolving appropriate intellectual property protection systems suitable for both developed and developing countries.

Addressing the issue of environmental safety in 1989 FAO, the International Office of Epizootics (OIE) and WHO jointly prepared the international requirements for vaccinia vector rinderpest vaccine. A joint FAD/WHO Consultation on Strategies for Assessing the Safety of Foods Produced by Biotechnology was organized in 1990. Other joint activities in this area are planned. The Joint FAD/IAEA Division operates a number of regional and interregional coordinated research programmes dealing with a variety of biotechnology applications in crop breeding and nutrition, animal production and health was held in 1991. FAO has also been collaborating with CGIAR in several biotechnology-related matters. For instance, it has been a member of the CGIAR Task Force on Biotechnology (BIOTASK) in discussing CGIAR-wide issues on biotechnology, including IPR and biosafety. FAO also organized a workshop for African biotechnologists in collaboration with IITA in 1990.

Biotechnology in the CGIAR system

As countries all over the world confront the need to boost agricultural output, biotechnology which comprises techniques for using plants, animals and microbes to produce useful products or improve existing species - stands out as a promising tool. These techniques, ranging from simple tissue culture methods to advanced genetic and molecular manipulation of biological material, hold out the possibility of relieving some of the present tolerance of crops and animals to particular physical or environment stresses (e.g.

salinity and drought), increasing the resistance of plants and animals to pests and pathogens, and increasing nutrient efficiency. Biotechnology has been defined thus: "Biotechnology is... comprised of a continuum of technologies, ranging from the long-established, and widely used technologies, which are based on the commercial use of microbes and other living organisms, through to the strategic research on genetic engineering of plants and animals".

Exactly how much modern biotechnology can or cannot do - especially for the developing countries - is still an unanswered question. Up to this point, most biotechnology research has taken place in the industrialized countries, concentrating on application of interest to the developed world. But in recent years, developing countries have become more active in the field, as they wish to partake of research benefits. A major study in 1988-90 by the World Bank, the International Service for National Agricultural Research (ISNAR), and the Australian Government, concluded that there were many potential benefits from integrating modern biotechnology with conventional agricultural research. This could be done by focusing public sector investments on the "orphan commodities" - crops traditionally of less interest to the industrialized countries but of great importance to the vast numbers of resource-poor producers and consumers in the developing world. The Consultative Group on International Agricultural Research (CGIAR) is one research group which is tackling these challenges.

Goals and Strategies of CGIAR

CGIAR was established in 1971 to support the works of several international agricultural research centres (IARCs). Today the CGIAR system consists of 18 IARCs, with support being received from some 40 donors. The CGIAR mission statement, as revised in 1991, states: "Through international research and related activities, and in partnership with national research systems, to contribute to sustainable improvements in the productivity of agriculture, forestry and fisheries in developing countries in ways that enhance the nutrition and wellbeing, especially of low-income people".

The ultimate aims are improved nutrition and economic well-being for low-income people, including women, landless labourers and poor producers and consumers in both rural and urban areas. Research should contribute to self-reliance by increasing the purchasing power of the poor through lower costs and prices and through greater equity in income distribution. It should also contribute to the quality of plant and animal products, to sustainability and stability in their supply, and to the prevention of environmental degradation through improved resource management.

CGIAR is complex and entails a comprehensive programme of research and related activities. Its goals are:

— effective management and conservation of natural resources (i.e. land, water, forests and germplasm) for sustainable production;

— improved productivity of high-priority crops, livestock, trees and fish, and their integration into sustainable production systems;

— improved utilization of crop, livestock, tree and fish products in both rural and urban areas through post-harvest technology;

— progress towards equity (including gender equity) as well as improved diets, nutrition and family welfare, through better understanding of human linkages between production and consumption;

— appropriate policies for the increased productivity of crops, livestock, trees and fish, and for the sustainable use of natural resources;

— strengthened human resources and institutions for greater research capacity in developing countries' research systems.

The system's involvement in each of these areas varies greatly. In some, such as crop and livestock production, the centres play a major role. In other areas, the CG system is primarily a catalyst, stimulating and supporting research at other institutions. And in other areas such as commodity conversion and utilization, the system integrates the work of other leading institutions into the results of the centre programmes. A combination of all of these approaches is used by the CG system in the area of biotechnology research.

It has been recommended that IARCs consider undertaking four initiatives in biotechnology to identify and transfer high-priority technologies to developing countries, explore opportunities for closer links with private industry, establish institutional biosafety committees, and establish a standing group of experts to deal with the role of biotechnology in world agriculture.

Biotechnology Strategy

A balance between biotechnology and conventional research is crucial to CGIAR. The strategy of CGIAR crop improvement centres is to use relevant, tested biotechnology techniques for more efficient execution of their research agendas.

The centres are interested in developing new tools for their own work as well as for their national programme partners. In some cases, new diagnostic, analytical, or plant breeding tools are developed by the centres to improve their own research capability in meeting critical international needs. In other cases, new tools are made and adapted specifically for national programme research efforts. A well-established linkage strategy of IARCs, that of using cooperation to pass on useful research techniques to national programmes, will also be a way forward in biotechnology, particularly for smaller NARS. One route may be to link IARCs with developing-country universities as future centres of strategic research. As skills mature, the universities can then exploit IARC connections to begin their own interactions with advanced institutions and companies in the industrial economies.

The "in-house" management of biotechnology in IARCs is rooted in this strategy and is still in a formative stage. Centres each have one, two or sometimes three scientists who act as coordinators of biotechnology research efforts. They listen to programme needs in demand and supply consideration together with the help of senior managers who deal with increasing concerns regarding contractual, legal and safety matters. In some centres, individuals have been the catalysts in making biotechnology research operational. Several of the larger centres have established separate biotechnology units.

Biotechnology Research

Biotechnology research in CGIAR and its potential applications were first discussed by the Group in 1981 (CGIAR, 1981). At that time, with the exception of the research of the International Laboratory for Research on Animal Diseases (ILRAD) which was well under way, mostly tissue culture techniques were being used by CGIAR scientists for multiplication and production of disease-free plant material, and genetic engineering techniques were discussed in a futuristic sense. Still, it was predicted that genetic engineering would not replace traditional plant breeding techniques, but would provide new tools for breeders. As science progressed, an overview was presented to a CGIAR mid-term meeting on the activities of biotechnology research in CGIAR in 1985; this overview was updated in 1988. By then it was clear that "biotechnology brings new tools, new ideas and new approaches to agricultural research. In most cases, biotechnology will not supplant what, for want of a better term, might be called conventional agricultural research. Rather, biotechnology will be used in conjunction with traditional plant breeding and other agricultural research to help provide new information and tools to solve problems."

Biotechnology includes both scientific advances and technologies that can be developed from those advances. For an advance to become useful, it must be packaged in a usable way. For example, hybrid maize is a technological delivery package for scientific advances that emanated from a new understanding of basic genetic principles, leading to the development of hybrid seeds that farmers could plant, to their benefit. An understanding of the relationship of science and technology can help us to see the potential benefits of biotechnology in the international agricultural research centres and their national programme partners.

CGIAR believes it can help in two fundamental ways: by providing a bridge for the flow of needed information and germplasm (seed and plant material) between developed and developing countries, and by ensuring that the agricultural needs of the developing countries are not lost in the total research effort, given that there would be little money to be made by private firms in the improvement of orphan commodities or in many technologies designed for resource-poor farmers.

Most IARCs that deal with crop or livestock improvement are involved, or are likely to become involved, in biotechnology research. This includes cellular and molecular biology and the new techniques derived from these to improve the genetic makeup and management of crops and animals. However, few IARCs will develop new basic knowledge in support of biotechnology, but rather will become involved in finding applications for new scientific advances. Thus IARCs are dependent upon scientific advances and resulting technologies in efforts to adapt technologies to developing country problems. In a sense, IARCs are toolmakers in biotechnology, using basic scientific knowledge to develop useful technologies in crop and animal research.

If IARCs are to play an important role in adapting biotechnology tools to the needs of developing country agriculture, new methods for acquiring technologies must be considered. It is true that commerce drives many of the applications of genetic engineering. Not only must the centres themselves be able to negotiate creatively with the private sector, but also appropriate research institutions in the developing countries may need to do so. Several developing countries' research institutions have already recognized the need to apply state-of-the-art technologies to their ongoing research efforts.

At present, the scale of CGIAR biotechnology research is modest - especially compared with that of the private sector - and probably will remain so. In 1990, for example, CGIAR plant and animal biotechnology research totalized US$14.5 million. This was less than 5 percent of the total CGIAR budget and only a traction of the US$55 million spent on agricultural biotechnology by one US corporation, the Monsanto Chemical Company. Half of the CGIAR outlay on biotechnology went to plant research, distributed across nine international centres working on 15 crops. The other half went to animal research, primarily at ILRAD in Nairobi. This centre conducts state-of-the-art molecular biology research for the control of two major livestock diseases: tick-transmitted East Coast Fever (theileriosis) and tsetse fly-transmitted African sleeping sickness (trypanosomiasis).

Role of IARCs in Advanced Research Networks

Many international centres play a crucial role in advanced networks that link centres with advanced basic science institutions. Such networks help to target research, develop methodologies, establish new research channels, expand awareness of centre-based research, and provide a means to acquire new technologies. The centres gain because they can link their own programmes with those of advanced and developing country institutions on problems of mutual interest. Collaborating advanced institutions benefit because they can gain an understanding of major developing country problems that might benefit from biotechnology approaches, as well as learn more about the genetic resource collections held by the centres.

The oldest such network, as far as the centres are concerned, is the Rockefeller Foundation-supported network on biotechnology in rice, established in 1985. This network links advanced laboratories in Europe, the United States and elsewhere, with the international Rice Research Institute (IRRI) and the International Centre for Tropical Agriculture (CIAT). Molecular maps are being developed to allow breeders to determine whether an individual rice plant contains known genes. Many potentially useful genes have been cloned from rice, including the gene "oryzacystatin", an inhibitor of the digestive enzymes of insect pests.

The formation of the Cassava Biotechnology Network in 1988 at CIAT, involving scientists in Latin America, Africa, Europe and the United States, brought new techniques to difficult research problems in cassava. A major effort involves studying the biochemistry and genetics of cyanogenesis (the formation of HCN, hydrocyanic acid, in plants). Social scientists will assist with farming systems studies in areas where cyanogenic cassava is grown. Findings will help define research objectives: whether to eliminate cyanide throughout the plant, or to increase specific enzyme activity to reduce the presence of cyanogenic compounds. A further goal is the production of true seed as a joint venture between the International Institute of Tropical Agriculture (IITA) and CIAT.

The International Potato Centre (CIP) and the International Centre for Maize and Wheat Improvement (CIMMYT) are also effectively using networks. CIP, which works with potato and sweet potato, has established an extensive network of collaborating institutions in the United States, China, Israel and Europe for activities in genetic engineering and RFLP analysis. Efforts such as this have enabled CIP to obtain access to scientific expertise in a cost-effective manner. CIMMYT was involved in establishing and participating in a maize RFLP network under the EUREKA funding and guidelines of Europe. This network consists of public and private institutions in France, Germany, the Netherlands and Italy. Their goals are to utilize molecular markers (RFLPs) to complement the genetic map of maize; to fingerprint European and CIMMYT maize inbred germplasm; to identify (map) Quantitative Trait Loci (QTLs) for important agronomic characters; and to develop strategies for using marker-assisted selection of breeding programme. CIMMYT also participates in the International Triticeae Mapping Initiative (ITMI), a network of scientists interested in mapping the wheat and related genomes.

In addition to the research networks, ISNAR established an Intermediate Biotechnology Service (IBS) in early 1993, to serve as an advisory network on management and policy aspects of biotechnology research in developing countries.

CIAT

Biotechnology research began in the period from 1980 to 1984, with emphasis on tissue culture for conserving cassava germplasm and accelerating rice germplasm improvement. In 1985 the Biotechnology Research Unit (BRU) was established. BRU focused further on critical challenges in CIAT's crop germplasm and expanded to include the study of selected microorganisms and modern biochemical and molecular techniques. Research is carried out through special project funding and includes:

— Pilot projects with IPGRI (the former IBPGR) on an in vitro genebank and cyropreservation. This work was completed in 1991.

— Gene tagging and mapping of Rice "Hoja Blanca" Virus (RHBV) and rice blast resistant genes (funded by the Rockefeller Foundation, this project is ongoing).

— Construction of a cassava molecular map, to be completed in 1994 (RF funding).

— Construction of a molecular map of tepary beans to be finished in 1994, funded by Belgian AGDC.

— Operations and coordination of the Cassava Biotechnology Network (CBN), 1992 - 1997, funded by Dutch DGIS.

CIAT also established several collaborative research projects with developed country institutions in the United States and Europe to work on, inter alias, genomic studies in cassava, rice biotechnology and molecular markers for evolutionary studies in common bean.

Additional research includes the characterization of mechanisms involved in resistance and tolerance to biotic and abiotic stress in plants. This work can be categorized as

(1) the characterization of resistance mechanisms to pests and pathogens (e.g. resistance to bruchids, a major pest in beans, and developing the molecular bases for co-evolution in bean pests);

(2) characterization of physiological and biochemical processes in plants and bacteria (e.g. drought and heat tolerance in cassava, aluminium tolerance to acid soils, and cassava starch fermentation); and

(3) development of methodologies. As technologies are developed, CIAT shifts activities from BRU into the appropriate research programme, which then continues the work. Since 1988 six major projects have been handled in this manner.

In addition to research efforts, CIAT has trained nearly 90 people from 20 countries in Latin America, Asia and Africa in tissue culture and the biochemical and molecular characterization of genetic resources. Future training will emphasize graduate studies and in-service research, linked as much as possible to national research programmes.

CIMMYT

Since the early 1980s, CIMMYT has utilized various tools of biotechnology in wheat and maize improvement. They have included the use of tissue culture for producing the inter genetic hybrid triticale, somaclonal selection for salt tolerance in wheat, and the enhancement of recombination in the wide-hybridization of wheat with wild relatives, and maize with Tripsacum. Enzyme-Linked Innunoserbent Assay (ELISA) assays were and are being used in the routine diagnosis of several viral diseases, including barley yellow dwarf virus (BYDV). Various protein and isozyme markers have been used in detecting chromosomes of wheat in evaluating protein quality of wheat varieties.

With the completion of its new applied biotechnology facilities in 1989, CIMMYT's efforts in biotechnology were enhanced to include the use of molecular marker technology for location and manipulating the genes involved in traits of agronomic importance. These efforts are the focus of the Applied Molecular Genetics Laboratory. More recently, the Tissue Culture Laboratory was established in order to take advantage of recent reports of successful maize and wheat transformation. All activities in biotechnology seek to enhance the breeding programme, through either more efficient selection techniques or enhanced germplasm. Technology evaluation and adaptation for use at CIMMYT and in its client countries, training and technology transfer and collaborative research, all play a part in these efforts.

The Applied Molecular Genetics Laboratory has concentrated in the last few years on establishing efficient protocols for the use of marker technologies in wheat and maize. There are four aspects of protocol development:

- use of non-radioactive DNA hybridization techniques to provide a safer working environment;
- adaptation of RFLP and other marker techniques for large-scale analysis;
- optimization of cost-effective experimental strategies; and
- development of software for fingerprinting databases, molecular data capture and verification and cost analysis.

Current collaborative research in maize includes the evaluation of the genetic diversity of tropical, subtropical, highland and African germplasm. The current database contains molecular fingerprints for over 100 lines. This information has enabled CIMMYT's breeders to understand better the heterotic groupings and to develop more appropriate testers for germplasm improvement. A major study involves the genetic base of resistance to multiple corn borers using RFLP genotyping and field evaluations of populations segregating for resistance. Results indicate resistance is polygenic and gene action is primarily additive. It is expected that RFLP markers will facilitate the incorporation of resistance into elite germplasm, increasing breeding efficiency and reducing costs.

The application of marker technologies in wheat is more complex, because of its larger genome and its polyploid nature. There has been progress in pilot mapping studies using large-scale RFLP analysis for durable leaf rust resistance and resistance to bacterial leaf blight. Projects are being developed to widen the genetic base of modern wheat through marker-assisted introgression of desired traits from wild bread wheat relatives (e.g. resistance to viruses, fungi and bacteria; and tolerance to abiotic stresses such as salinity, drought and aluminium). Molecular markers are expected to become routine breeding tools at CIMMYT; this will require a "service-oriented" laboratory to handle the expected large numbers of routine molecular analyses.

The Tissue Culture Laboratory seeks to enhance the resistance of tropical maize to major insect pests through transformation which offers breeders a new tool for broadening the genetic composition of germplasm. Several major tropical and subtropical CIMMYT lines are being evaluated for embryogenic callus formation and regeneration potential. Also under way are insect bioassays of partially purified toxins from Bacillus thuriengensis (Bt) strains collected in Latin America in collaboration with the Mexican plant molecular biology institute, CINVESTAV, and the EMBRAPA maize breeding station in Sete Lagoas, Brazil. Efforts in maize transformation as well as collaboration in several developing and developed country institutions will soon be enhanced, with the recent approval of a special project under UNDP funding.

CIP

Recent work expands on the new Restriction Length Polymorphism (RFLP) map for potato produced by Cornell University, in collaboration with CIP. A total of 220 markers are now available. Randomly Amplified Polymorphic

DNA (RAPD) markers, generated by the Polymerase Chain Reaction (PCR), have been adapted to analyze DNA variation in tuberbearing Solanum, demonstrating the RAPD markers are suitable for large-scale assessment of the genetic diversity of potato populations.

Transformation through the insertion of antibacterial protein genes and the coat-protein gene for Potato Leaf Roll Virus (PLRV) into potato clones is now possible. Two genotypes, Desiree and 86007, were used for transformation with Agrobacterium tumefaciens and A. rhizogenes as vectors for gene coding for antibacterial proteins.

CIP's virus detection work continues to provide antisera to the major potato viruses, and a new PLRV antiserum is being prepared. Virus detection kits for potato viruses are a widely used CIP product. More than 750 000 virus samples were evaluated through the distribution of antisera, Das-ELISA, and NCM-ELISA kits. There is a great demand for this technology in all countries with severe Sweet Potato Weevil (SPW) problems since progress has been slow in developing SPW-resistant cultivars. The recent discovery of Bt strains effective in controlling SPW has opened new avenues for genetically engineered sweet potatoes with the Bt-toxin gene. In the near future transgenic sweet potatoes with resistance to SPW may become available.

To develop transgenic potatoes with resistance to tuber moth, CIP is collaborating with the Comitato nazionale per la ricerca e per lo sviluppo dell'energia nucleare e delle energie alternative (ENEA) in Italy, Michigan State University (MSU), and Plant Genetic Systems (PGS) of Belgium. ENEA has focused on the development of two DNA sequences similar to other known Bt sequences. These were used for transforming potato plants by Agrobacterium infection. Collaborative work between MSU and PGS has produced several potato cultivars with resistance to PTM. Several CIP advanced clones have now been selected for engineering Bt genes with resistance to PTM. Other genes of interest for transformation efforts include the Cowpea Trypsin Inhibitor Gene (CPTI).

ICRISAT

ICRISAT scientists use biotechnology tools to overcome crop production constraints, but only where conventional techniques cannot be applied, or where biotechnology is more efficient or cost-effective. ICRISAT adapts proven biotechnology techniques to its research.

In cereal biotechnology research, ICRISAT aims to use molecular markers in crop improvement. In collaboration with the University of Milan, a RFLP map of sorghum is being developed using maize probes. Similarly, with the collaboration of scientists from the United Kingdom, a RFLP map of pearl millet is nearing completion. Refined molecular maps for both crops will enable researchers to make rapid progress in their improvement. ICRISAT participates actively in global efforts to use molecular markers in crop improvement.

Some transformation work is also being pursued. Whole plants were regenerated from various organs or single cells of pearl millet, and from shoot apices of sorghum. Future work will build upon collaboration with public institutions in the United States and elsewhere, particularly in sorghum.

In collaboration with the Scottish Crop Research Institute (SCRI), United Kingdom, a molecular map of groundnut is being developed to identify markers for agronomic traits. This will help identify the coat protein gene of the Indian peanut clump virus and make plasmid constructs suitable for the transformation of groundnut. Research in chickpea and pigeonpea has emphasized reliable tissue culture and plant regeneration techniques for use in wide crossing and transformation. Interspecific hybridization of chickpea (Cicer arietinum) and a wild Cicer spp. is in progress. Target genes are those conferring resistance to insects and to fungal pathogens. For pigeonpea, wild species hold many desirable characters not present in the crop. An interspecific hybridization programme involves crossing Cajanus cajan with Cajanus platycarpus, and overcoming current barriers to hybridization.

IRRI

As mentioned earlier, IRRI's biotechnology work has been supported by the Rockefeller Foundation International Programme in Rice Biotechnology, established in 1985. This network serves as a prototype for successful interactions between developed and developing country institutions, and includes the participation of IRRI, CIAT, IFPRI and ISNAR. Of the total amount invested or committed to the programme by the Rockefeller Foundation since 1985 over US$48 million - slightly over US$22.5 million have gone to institutions in North America, Europe and Japan; nearly US$19 million to developing countries - most of the funds being spent on 130 fellowships; US$ 1 million towards information dissemination; and slightly over US$6 million to international agricultural research centres. Of the funds going to IARCs, US$5 million supported 12 biological research projects, and US$ 1 million supported six social science research projects.

The rice biotechnology programme aims to strengthen the capacity of research institutions in developing countries to assume increased responsibility for further development and application of rice biotechnology. IRRI, in collaboration with developed country laboratories, has worked to bring about some of the major highlights of the programme, including the following:

— a molecular genetic map of rice with over 600 markers has been developed and distributed around the world and is now being used to tag important genes;

— rice transformation has been achieved for major tropical (indica) rice-breeding lines such as IR72. Potentially useful transgenes, such as those for virus and insect resistance, have been transferred to rice and are currently being tested;

— DNA markers of rice pests and pathogens have been used to determine their genetic diversity and structure; markers have helped breeders to identify resistant germplasm and to breed for durable resistance;

— developing country scientists have participated in all research projects.

Information and technologies generated are passed to developing country institutions for further research and implementation.

ILRAD

ILRAD's work represents approximately half of CGIAR funds invested in biotechnology research. Recent accomplishments include joint work with the University of California at Berkeley which led to a completed map of the entire genetic material of Theileria parva, an important parasite of domestic African livestock, which causes the disease known as East Coast Fever. This restriction map is the first complete map made of the genome of a protozoan parasite. Techniques involving modern molecular biology, biotechnology and genetic manipulation were used in the mapping exercise. The parasite genome map will be used to monitor the occurrence of genetic recombination in the parasite's life cycle, which involves the generation of a wide variety of parasite strains. Knowledge of the basis of this diversity will help researchers develop a vaccine that will work broadly and effectively against all strains.

The theileriosis programme continues to develop and refine antibody-based ELISA systems and DNA probes for precise identification of tickborne disease organisms. In this effort ILRAD works closely with laboratories in Australia, United States and Latin America. ILRAD has identified, characterized, cloned and expressed a surface antigen of the sporozoite stage of T. parva. The recombinant protein when combined with adjuvant is capable of inducing protective ability has also been identified, characterized, cloned and expressed, and its immunizing potential in cattle is now being tested. These antigens represent the basis for the development of new, genetically engineered vaccines against East Coast Fever.

The trypanosomiasis research programme has developed highly sensitive tests (monoclonal-antibody-based ELISAs) for distinguishing trypanosome species. The reliability of this work was validated in 1991 by researchers in eight laboratories in Africa whose work was supported by FAO and the International Atomic Energy Agency (IAEA) in Vienna. The utility of these tests in diagnosing human trypanosomiasis (sleeping sickness) is now being tested with the World Health Organization in Geneva. ILRAD and its partners in a network of laboratories in the United Kingdom, Switzerland, Israel, the United States and Australia, are engaged in developing a physical map of the bovine genome. In this collaboration, ILRAD provides a major genetic resource by breeding F1 and F2 families derived from crosses of trypanosusceptible and trypanotolerant parents and determining their trypanotolerant status. ILRAD has also developed technologies and primers to define cattle populations and subpopulations, known as RAPDs, which will be highly valuable in the description and conservation of global bovine genetic resources.

Training and Institution Building

An important way to diffuse research technology is through training and institution building. CGIAR has an illustrious record in research training to help strengthen developing country capabilities. Between 1985 and 1989, CGIAR-supported centres conducted training courses, most of them lasting from two to four weeks, to an estimated 25,000 developing country researchers and scientists in all areas of crop and livestock improvement. In addition, research fellowships in biotechnology have been offered at several centres.

Most IARCs are hesitant to promote biotechnology training in countries without significant plant breeding programme and without appropriate laboratory facilities. Experience shows that researchers returning to a national system, after receiving training from a developed country institution but with little or no means of mobilizing their skills, become frustrated and often feel "cut off" from scientific advances.

On the institutional side the centres are increasingly being asked by national agencies to help set up creative programmes that often have a biotechnology component. In 1986, for example, the Andean Development Bank (CAF - Corporación Andina de Fomento) asked IARCs in the Latin American region to assist in some of the activities of its recently created biotechnology programme. In response, CIP began work with CAF on projects in five Andean countries (Peru, Bolivia, Ecuador, Colombia and Venezuela), which range from using tissue culture for disease-free material, to in vitro potato seed production, to facilitating the transport of planting materials to remote areas.

Information Dissemination

As research advances are made in biotechnology, the importance of information flow becomes essential. IARCs help NARS by providing technical information which may be otherwise difficult to obtain. Information resources currently available include Ag-biotech news and information, published by CAB International. The CGIAR Biotechnology Task Force (Biotask) was formed in 1988 to help raise the awareness of biotechnology in all elements of the CGIAR system, including lARCs, donors, TAC and national partners. In addition to providing a forum for discussion on issues of biotechnology, other activities were undertaken, such as assistance by the Dutch Government in distributing the ag-biotech journal to 120 developing country libraries. Biotask ended its work in 1992.

In regard to information needs, in 1990 the Directors-General of the CGIAR centres stated:

> "The fullest application of biotechnology may be constrained by non-technical issues such as biosafety regulations, intellectual property management and public awareness. The dissemination of balanced, accurate information concerning biotechnological activities carried out in the Centres and elsewhere in the scientific community in the service of agriculture would ease such constraints".

Current and Future Policy Considerations

Biosafety

Biotechnology regulations in the developing world are complicated, partly because of a diversity of players in biotechnology research and testing, including IARCs, their donors, developing country governments and research institutions, research institutions and private industry of industrialized nations, and various environmental and public advocacy concerns. Each has its own agenda and definition of what constitutes safe and appropriate release.

All biotechnology programme must abide by the legally established regulations of the country where the research is being conducted. Currently, if regulations exist at all, they vary widely from country to country. If regulations are not in place, international guidelines to ensure adequate safeguards are followed. A common first step is the successful testing of regulated organisms in industrialized countries or in countries with established regulatory and approval mechanisms; this helps clear the way for subsequent testing in developing countries. However, care must be taken when "transposing" the regulations of one country directly to another as the different ecologies, crops, quarantine regulations, etc. vary greatly from one site to another; each site must make its own policy and cost/benefit analyses.

When applying new technologies, IARCs must deal with many issues including scientific, technical, environmental, regulatory and policy dimensions of research. Several centres have established special committees to prepare for the testing of such genetically improved material. IARCs follow international guidelines set by OECD, and work closely with host countries in determining proper risk analysis for field testing.

Plant Genetic Resources and Intellectual Property Management

The need to preserve future biodiversity has been, and continues to be, a top priority for CGIAR. Over the years, several of the centres have assembled global germplasm collections for their mandated crops, often using the material to re-establish individual country collections (e.g. because of storage failures or losses caused by political unrest). The former International Board for Plant Genetic Resources (IBPGR) now the International Plant Genetic Resources Institute (IPGRI) - was established in 1974 for the purpose of promoting germplasm utilization and conservation.

CGIAR has always operated an "open door" germplasm policy - meaning that all clients, whether public or private, in developed or developing countries, can enjoy access to the germplasm. But the involvement of the private sector in biotechnology research has raised some questions about both access of developing countries to new biotechnologies and the extent to which research will benefit resource-poor farmers. Developing countries see themselves as faced with two possible scenarios, either of which may be entirely acceptable. They can protect intellectual property rights, and thereby obtain access to the latest crop technologies which in turn may lead to higher seed costs. Or they can choose to forego IPR protection, perhaps to save foreign exchange and attempt to protect farmers' incomes, but risk being left behind in the technology race. Are alternative scenarios possible?

Intellectual property considerations are not new in CGIAR. ILRAD has long sought vaccines for its two target diseases, trypanosomiasis and theileriosis. ILRAD realized that to develop a vaccine and make it available to African farmers paradoxically it would have to file for patent protection. Such protection was considered essential to satisfy private sector partners who otherwise would be unwilling to provide the investment funds necessary to develop and produce a vaccine and bring it to market. Other centres have obtained patents on discoveries, such as small farm machines developed at IRRI, but such patents were taken for defensive purposes. CGIAR does not seek financial gain from discoveries made at IARCs. When patents are taken, they are used by IARCs themselves or transferred to their partners to ensure that scientific advances are utilized in developing countries.

In 1982 CGIAR began to review the implications of intellectual property protection, beginning with a report by TAC on the role of plant breeders' rights, and culminating in a recent discussion paper on the subject of plant genetic resources, biosafety and intellectual property rights. The latter paper is currently under review both within and outside the CGIAR system. At the mid-term meeting of CGIAR in Istanbul, in May 1992, the members of CGIAR agreed to a CGIAR Working Document on Genetic Resources and Intellectual Property which stated:

> "The CGIAR reaffirms that genetic resources maintained in the genebanks of the centres are held in trust for the world community. Material from the genebanks at the centres will continue to be freely available, in accordance with the 1989 CGIAR Policy on Plant Genetic Resources... Centres do not

seek intellectual property protection unless it is absolutely necessary to ensure access by developing countries to new technologies and products. The centres will not seek intellectual property protection as a source of operating funds. Should exceptional cases arise where a centre might receive a financial return, an appropriate means will be used to ensure that such funds are used for the conservation of genetic resources and related research".

Future outlook

Future CGIAR efforts in biotechnology are likely to emphasize genetic characterization of germplasm materials of important crop plants and animals, improved diagnostic tools, and enhanced genetic improvement of important crops and livestock. Partnerships will continue to be sought with advanced institutions and developing country institutions to help solve important common problems, in particular those of developing countries. As now, IARCs will largely be consumers and not originators of developments in basic sciences, but will stand among the leaders in developing tools and technologies that can be used in developing countries. Furthermore, IARCs will play a bridging role in technology transfer and a leading role in helping to train developing country scientists.

Crop Biotechnology in the Asia-Pacific region

Although almost all countries in the region have biotechnology programmes, the level of advancement and involvement varies widely. The three developed countries of the region, namely, Australia, Japan and New Zealand, and particularly the first two, are frontline countries in the world in R & D of this new and fast developing field As regards the developing countries, moderate to comprehensive biotechnology R & D programmes are taking place in China, India, Indonesia, Malaysia, Pakistan, the Philippines, the Republic of Korea and Thailand. Of the remaining developing countries, Bangladesh, the Democratic People's Republic of Korea, Iran, Nepal, Sri Lanka and Viet Nam are striving to establish appropriate biotechnology programmes. Several of the Pacific island countries have also developed in vitro culture facilities for taro and other selected crops essentially for micropropagation, freeing of planting materials of viruses and quarantine purposes. Fiji has also made considerable progress in somaclonal selection and in vitro culture of sugar cane.

Biotechnology in Crop Improvement and Production

In vitro culture for micropropagation of virus-free materials

Vitroplants of potato, sweet potato, Allium species, taro, strawberry, sugar cane, several ornamentals such as chrysanthemum, carnation, Lilium species and gerbera, and fruit plants such as Citrus species, papaya, apple, sweet cherry, pear and grapes freed of or free from viruses are being commercially exploited in several countries in the region, particularly in China, India, Indonesia, Japan, Malaysia, the Philippines, the Republic of Korea and Thailand. Hundreds of thousands of hectares are planted to virus-free vitroplants of potato in the Region with more than doubled productivity, especially in China and the Republic of Korea. For instance, in the Republic of Korea, the widespread use of virus-free planting material produced through in vitro culture techniques increased the national potato yield from 11.9 tonnes/ha in 1980 to 20.3 tonnes/ha in 1986. Subsequently, an in vitro tuberization system was established and became an integral component of the potato seed industry in the country. Production efficiency of the technique is exceptionally high. One of the main problems encountered in this approach is the chance of reinfection during the propagation of virus-free plants. To overcome and manage this problem, simplified diagnostic methods have been developed and are being used to detect viruses and viroids.

In vitro culture for micropropagation

Besides the above, vitroplants are being routinely mass micropropagated for commercial purposes in the case of orchids, other flowers and ornamental plants throughout the region, especially in Southeast Asian countries, oil palm (Malaysia, India, Indonesia, Papua New Guinea), date palm (Iran and likely to be taken up in India and Pakistan), embryo culture of soft coconuts (Macapuno) (the Philippines and Thailand), cardamom, eucalyptus (China and India), Chinese fir (China), rattan (Indonesia, Malaysia, the Philippines), and seed sterile triploid water melons (Japan). Somatic embryogenesis is also being commercially applied. Synthetic seeds of rice and vegetables have been developed by the private sector in Japan. Triploid rubber clones of rubber produced through somatic embryogenesis have outyielded diploid standard clones by about 20 percent in China. The country has also standardized techniques for mass propagation of sugar cane seedlings using embryogenic cell lines and multishoot propagation systems.

To facilitate reforestation and promote social forestry and agroforestry, India has established two tissue culture pilot plants, one at the National Chemical Laboratory, Pune, and the other at the Tata Energy Research Institute, New Delhi (a private organization) with production capacities of five to ten million propagules/seedlings of elite/plus trees of several important species, such as Eucalyptus tereticornis, E. camadulensis, Tectona grandis, Dendrocolamus strictus, Populus deltoides, Bambusa vulgaris and B. tulda. Micropropagation of teak, rattan, eucalyptus, bamboo and other tree species has been adopted in several other countries of the region.

Doubled haploid breeding

The region has played a pioneering role in this field. Indian scientists were first to produce haploids from another culture in the mid-1960s. Haploid induction through anther culture has been used most extensively by Chinese scientists for breeding of not only rice but also of several other crops. The technique was used for: (i) production of new varieties; (ii) production of inbreds for heterosis breeding; (iii) selection of stable resistant lines against biotic and abiotic stresses; and (iv) purification of male sterile lines. One of the most important advantages of the use of haploids is to reduce the time required to develop and release new varieties. For instance, while the conventional pedigree method of breeding and releasing a pure line rice variety requires about 12 years, the haploid method requires only about five years.

Chinese scientists have succeeded in inducing haploids in about 40 different plant species. They were the first to produce haploids through in vitro culture in wheat, maize, sugar cane, soybean, rubber, grapes and apple. In China, several new varieties of rice, wheat, tobacco, and hot and sweet pepper possessing high yield, superior quality, sometimes tolerance to abiotic stresses such as cold, sometimes early maturity, and resistance to diseases have been released through the use of haploids. The most popular of these varieties are Xin-Xin, Hua-Hau-Zao, Zhong-Hua 8, 9, 10 and 11 of rice; Jing-Hua 1, 3 and 5 of wheat; Haihua 1, 3, 19 and 29 of hot and sweet pepper; and Dan-Yu 1, 2 and 3 of tobacco. These varieties occupied about 1.5 million hectares in 1990, of which the rice and wheat varieties accounted for 800 000 and 650 000 ha, respectively.

Other countries have also used doubled haploid breeding methods for crop improvement. In India, rice lines R4 and R10 derived through this method outyielded the check by 15 to 49 percent in trials conducted during 1984-87. The Republic of Korea released two rice varieties derived from anther culture haploid technique and generally screens annually about 6 000 anther-derived rice lines. The technique has also been effective in heterosis breeding of Chinese cabbage. In Japan, several successful varieties have been developed through this technique. Among these are rice varieties Joiku N. 394, Hirohikari, Hirohonami, AC No. 1 and Kibinohana, which are tolerant to cold temperature and are good in taste. A broccoli variety, "Three Main", possessing cold resistance and uniform shape, and a cabbage variety, "Orange Queen", with an orange colour, which are quite popular, were developed by the anther culture method.

In vitro selection and somaclonal variation

Australian scientists were the leaders in the field of somaclonal variation, demonstrating the efficacy of the approach in improvement of wheat, sugar cane and other crops. Somaclonal variants and lines selected through selection pressure exerted during the culture stage have been released as commercial varieties in several countries in the region. For instance, in Japan, the 1991 list of registered crop varieties included two varieties of Lilium and one each of melon, Gerbera, Cymbidium, statice and strawberry derived through somaclonal variation. In China, somaclonal mutants of rice possessing resistance to bacterial blight and resistance to AEC coupled with higher contents of Iysine, methionine, isoleucine, serine and tyrosine than the parental lines have been produced. Salt tolerant somaclones of rice, wheat and tobacco were also isolated. Short stature and early maturing somaclonal variants of rice have been isolated in several laboratories in China. India and other countries have also produced somaclonal variants and are using them directly as new lines or in breeding programmes. In Fiji, tissue culture techniques have been used extensively for isolating sugar cane clones resistant to Fiji disease of sugar cane. Thailand isolated improved varieties of banana and chrysanthemum using this technique.

Somaclonal variation was exploited in a novel way under a joint programme of China and Australia to transfer from Thinopyrum intermedium resistance to Barley Yellow Dwarf Virus (BYDV) in wheat (Triticum aestivum). Single cell callus cultures from F1 embryos rescued were initiated and induced to form plants showing somaclonal variation which were then selected for BYDV resistance. Cytological analysis of the genotypes showing stable resistance revealed that chromosomal rearrangement of the chromosome carrying Thinopyrum introgressed segment had occurred during the tissue culture phase to confer the stability.

The Biotechnology Centre at the Indian Agricultural Research Institute (IARI) has standardized the protocols of plant regeneration of Brassica carinata and is isolating somaclonal variants suitable for Indian conditions. Useful somaclonal variants for earliness, plant height, maturity, etc. have been induced in B. juncea and B. napus.

In vitro techniques for plants provide systems analogous to the prokaryotic systems where variants can be selected and mutations can efficiently be induced and isolated at cellular level. By applying effective selection pressure on naturally variant or mutagenically treated cell cultures viz. use of saline media for screening salinity resistant/tolerant cell lines or use of pathogen toxins for isolating disease resistant cell lines, the efficacy and efficiency of both induction and identification of useful mutants are increased considerably. Using this technique, Chinese, Indian and Filipino scientists have isolated rice mutants tolerant to higher concentrations of salt. Mutants having higher protein and Iysine content in their seeds were also isolated. Using this approach, pathogen-resistant mutants of tobacco, rice, wheat, barley, sugar cane and maize have been isolated and used in breeding programmes in several countries. In Japan,, disease-resistant lines of rice, tomato and tobacco were isolated using this approach.

Embryo Culture and Distant Hybridization

The technique is being used in several Asian countries, particularly in the case of orchids, peanuts, cotton, cabbages, citrus and peaches. In India, promising recombinants have been obtained using embryo rescue techniques in distant crosses of cultivated Phaseolus, jute and peanut. In China, using embryo rescue and in vitro culture of a distant hybrid Triticum aestivum x Agropyron elongatum, a new wheat variety, Xiaoyan No. 6, was produced and has been grown widely giving additional wheat production. In Japan, three cultivars each of Citrus, Prunus, and Brassica species, and five cultivars of Lilium species were developed in recent years using the embryo rescue technique.

Cell fusion and somatic hybridization

Japan and China are particularly deeply involved in this work. During the past ten years or so, as listed by Nakajima, Japanese scientists have reported successful protoplast culture in more than 70 plant species. In China, plants regenerated from protoplast have been obtained in about 30 species, including vegetables, medicinal plants, legumes, and other economic crops as well as woody species such as poplar, elm and rubber trees. For the first time, Chinese scientists regenerated plants from monocot Polypogen fugax. With recent reports of success on protoplast culture and regeneration of wheat, it is now possible to have protoclones of most of the major food crops. However, in several cases, the regeneration frequency is low and should be improved. The Republic of Korea had produced cybrid lines of mushrooms which outyielded the best parents and checks by about 100 percent.

Successful regeneration of plants from cell fusion are reported from about 50 interspecific and intergeneric protoplast combinations in the region. Half of these are known from Japan. Using cell fusion techniques, scientists have developed novel citrus hybrid varieties, such as "Oretachi" (orange plus trifoliate orange), "Shuvel" (Satsuma mandarin plus navel orange), "Gravel" (grapefruit plus navel orange), "Murrel" (murcott plus navel orange) and "Yuvel" (Yuzu plus navel orange). Using asymmetric cell fusion techniques, Japanese scientists developed Ms-F 224, the first commercial tobacco male sterile line developed in the world bred by cell fusion. Such asymmetric male sterile lines for commercial exploitation of F, hybrids have been also bred in carrots, cabbages and eggplants.

In China, somatic hybrids between Nicotiana tabacum and N. rustica, N. tabacum and N. glauca were used for developing new commercial varieties of tobacco. Chinese scientists have also pioneered the pollen tube gene introduction technique for cotton and rice breeding. In this technique, after self-pollination, the desired exogenous DNA is introduced to the embryonic cells. Seeds which develop from such transformed embryos result into transformed plants, thus avoiding the need for protoplast fusion technique. Genes responsible for disease resistance and other traits have been transformed successfully in rice and cotton.

But this technique has low repeatability. In fact, the entire cell fusion technique has not been as successful in producing somatic hybrids as initially expected. With the availability of Agrobacterium-mediated and biolistics-based methods of gene transfer, the thrust on the protoplast fusion approach has somewhat slackened, although the asymmetric fusion method has special appeal for production and diversification of cytoplasmic male sterility and manipulation of other cytoplasmically controlled systems.

In vitro culture for secondary metabolites

Cultured plant cells retain their metabolic potential and can be used efficiently for producing useful secondary metabolites of commerce such as pharmaceuticals, dyes, food additives, sweeteners, flavours and taste enhancers, essential oils and aromatic products. Herbal medicines are popular in China, India and other Asian countries and are gaining importance in the West. Chinese scientists have been using tissue culture since the late 1970s for the production of ginseng sapanins from Panax ginseng. Diasgenin, a female contraceptive produced from in vitro cultured Dioscorea spp and a male contraceptive based on gossypol from tissue-cultured Gossypium spp are under extensive trials in China. In addition, tissue culture technique has been successfully used in the cultivation of Scopolia acutangula, Artemisia annua, and Rauwolfia yunnanesis. India is also using this technique for production of diasgenin and other medicinal and aromatic products.

Cell cultures can also be used as factories for bioconversion of intermediate compounds into more valuable products. Shikonin, an expensive compound obtained from the roots of Lithospermum erythrorhizon, has been used by the Japanese traditionally as a vegetable dye and in cosmetics and toiletries. However, due to overexploitation and other human activities, the plant has become almost extinct in Japan. To reduce their dependence on the import of this plant material, Japanese scientists have developed a tissue culture method for the commercial production of shikonin. Another example where tissue culture production of an industrial compound has reached commercial level is berberine from Coptis juponica. In tissue cultures the yield of high value compounds can be enhanced by feeding the cells with the intermediate compounds of their biosynthetic pathways (biotransformation), manipulation of culture conditions and selecting high yielding cell lines. With further refinements of techniques the bioreactor and fermenter based production of secondary metabolites could be rendered highly cost-effective and time-saving.

In vitro conservation of germplasm

Several plant species in the region, including a few commercial species, produce recalcitrant seeds and it is thus difficult to conserve them through seeds. Furthermore, some species are shy seed bearers and even fail to produce seeds. In addition, living collections of clonally propagated perennial crops face the problems of maintenance of heterozygous and heterogeneous populations, the long life cycle and large space required, and a high possibility of exposure to the threats of pests and diseases and other biotic and abiotic stresses. To circumvent these difficulties, in vitro conservation of vegetatively propagated and recalcitrant seed-producing species is being increasingly adopted as a complement to other methods of conservation. For short- to mediumterm storage (working and active collections), the slow growth method is used whereas for long-term storage (base collections) cryopreservation is the method adopted.

There is therefore a good case in the region for in vitro storage of several of the important plant species, and the countries involved are in fact building such facilities. Orchid germplasm is routinely maintained in vitro in India, the Philippines and Thailand. With assistance from the Department of Biotechnology, the National Bureau of Plant Genetic Resources (NBPGR), in India has established extensive in vitro storage facilities and is already storing germplasm collections of certain root and tuber crop species as well as a few fruit species. The bulk of the Indian potato collections at Central Potato Research Institute, Shimla, are conserved through tissue culture. China is also developing comprehensive facilities for in vitro conservation.

The Southeast Asian Banana Germplasm Bank, currently maintained as a living collection in Davao, the Philippines, has already been put under in vitro storage. The country is also maintaining some of its desired soft coconut germplasm through embryo culture. Malaysia is in the process of duplicating some of its ex situ living collections of oil palm and rubber as in vitro collections. Several of the Pacific Island countries are maintaining their taro collections in vitro because of the danger of the accessions being lost to viral and other diseases under field conditions. Thus, there is wide scope for using biotechnology for conservation and utilization of tropical plant germplasm.

Diagnostics

Besides the developed countries, several developing Asian countries are using monoclonal antibody techniques for identification of pathogens and for indexing materials. In China, more than 50 kinds of hybridoma strains have been constructed which secreted various kinds of monoclonal antibodies to viruses, bacteria, cancer, etc. India is applying monoclonal antibody techniques for the diagnosis and epidemiology of tungro virus of rice and other important viruses. The Central Potato Research Institute, Shimla, India, has been using monoclonal antibodies, cDNA and enzyme-linked immunosorbent assays (ELISA) for detecting and characterizing viruses of potatoes. Several other developing countries have also been using this technique coupled with ELISA tests to detect peanut viruses. Thailand has been using cDNA probes for detecting Tomato Yellow Leaf Curl Virus (TYLCV), Papaya Ringspot Virus (PRY) and mycoplasma causing sugar cane white leaf.

Recombinant DNA technology and transgenics

As regards the developed countries of the region, Australia and Japan have comprehensive programmes on recombinant DNA technology and production of transgenics for commercial exploitation. Australia was the first country in the world to have engineered and released a micro-organism for biological control of crown gall in plants. Australia's first field trial of transgenic plants took place in 1991, involving transgenic potato varieties resistant to leafroll virus, developed by the Commonwealth Scientific and Industrial Research Organization (CSIRO) Division of Plant Industry. During 1988-91, several laboratories in Japan, using electroporation or Ti or Ri, reported successful production of transgenics in Oryza saliva, Citrus sinensis, Cucumis melo, Lactuca saliva, Solanum tuberosum, Nicotiana tabacum, Morus alba, Actinidia chinensis, Atropa belledona. Brassica oleracea, Lyçopersicon esculentum, Vigna angularis and Vitis vinifera. For effective gene expression, promoter regions/genes have been identified for tissue-, age- and pathogen-specific expression of the genes under transfer. Stable and useful transgenics for protein quality and for viral resistance have been produced in rice, potato, tomato, melon and tobacco.

Molecular studies, gene tagging and mapping studies have been intensified in several laboratories in Japan. In 1991, a rice genome project, on the lines of the famous human genome project, was initiated. Restriction Fragment Length Polymorphs (RFLPs),, YAC clones or cosomid library and Several Sequence-Tagged sites (STSs) approaches are being deployed for the purpose. Cloned chromosome segments including useful genes are being analysed and characterized at molecular level. For identification and characterization of rice somatic chromosomes in in situ hybridization, imaging methods have been developed. Several countries in the region are using RFLP, RAPD and other techniques for gene tagging and creation of genome maps not only of major food crops but of topical industrial and fruit crops such as rubber, oil palm, durian and mangosteen. This approach is likely to increase the interaction between biotechnologists and plant breeders.

In China, protocols for production of transgenics for disease resistance in rice, tobacco, soybean and Brassica sp. have been standardized and transgenics are at various stages of development. Chinese scientists have attained considerable success in developing oilseed rape lines resistant to turnip mosaic virus through genetic engineering means. In India, transgenics are at various stages of production in cotton, rice, jute, Vigna aconitifolia and Brassica sp.. In the Republic of Korea, transformants were produced in tobacco, rice and tomato and work has been intensified for identification and isolation of useful genes, especially for disease resistance. The country is developing improved vectors for increasing the efficiency of transformants production. Malaysia is engaged in developing transgenics rubber and oilpalm, besides rice. In Pakistan, biotechnological manipulation of protein content and quality and resistance to Ascochyta blight in chickpea has been pursued in the last few years. Thailand, using the coat protein approach, has developed tomato and papaya lines resistant to TYLCV and PRV, respectively. Moreover, a mild strain of PRV has been developed for the mass preimmunization programme.

References

Ayyapan, S. 1988. Biotechnology and aquaculture, p. 7-8. *Asian Fisheries Society Indian Branch Newsletter*, October.

BIOTECH. 1990. *A decade and beyond - the National Institutes of Biotechnology and Applied Microbiology*. University of the Philippines at Los Baños, College, Laguna, the Philippines. 20pp.

Goodman, D.C. *From Farming to Biotechnology: A Theory of Agro-industrial Development*. Oxford: Blackwell. 1987.

Gianessi, L. P., C. S. Silvers, S. Sankula, and J. E. Carpenter, "Plant Biotechnology: Current and Potential Impact for Improving Pest Management in U.S. Agriculture, an Analysis of 40 Case Studies", Washington, D.C.: National Center for Food and Agricultural Policy, 2002.

Morgan, Sally. *Superfoods: Genetic Modification of Foods (Science at the Edge)*. Heinemann. 2003.

Bibliography

Ammann, K., *Biodiversity and Agricultural Biotechnology: A Review of the Impact of Agricultural Biotechnology on Biodiversity*, Bern, Switzerland: University of Bern Botanical Garden, 2003.

Atherton, K. T., ed., *Genetically Modified Crops: Assessing Safety*, London and New York: Taylor and Francis, 2002.

Ayyapan, S. 1988. Biotechnology and aquaculture, p. 7-8. *Asian Fisheries Society Indian Branch Newsletter*, October.

Bhat, S. F., and S. Srinivasan, "Molecular and genetic analyses of transgenic plants: considerations and approaches", *Plant Science,* 2002.

Bhojwani, S. S., Razdan M. K. *Plant tissue culture: theory and practice, Revised edition* Elsevier, 1996.

BIOTECH. 1990. *A decade and beyond - the National Institutes of Biotechnology and Applied Microbiology.* University of the Philippines at Los Baños, College, Laguna, the Philippines. 20pp.

British Medical Association. *The Impact of Genetic Modification on Agriculture, Food and Health.* BMJ Books. 1999.

Conger, B.V. *Cloning agricultural plants via in vitro techniques.* CRC Press Inc. 1981.

Conway, G. Genetically modified crops: risks and promise. *Conservation Ecology* 4(1): 2. 2000.

Conway, G., *The Doubly Green Revolution: Food for All in the Twenty-first Century*, Ithaca: Cornell University Press, 1997.

Cowan, C. W., and P. J. Watson, eds., *The Origins of Agriculture: An International Perspective*, Washington, D.C.: Smithsonian Institution Press, 1992.

Custers, R., ed., *Safety of Genetically Engineered Crops*, Flanders: Inter-university Institute for Biotechnology, VIB publication, 2001.

Debergh Zimmerman, (ed.) *Micropropagation: technology and application.* Kluwer Academic Publishers. 1991.

Domach, M. M. *Introduction to biomedical engineering.* Upper Saddle River: Pearson Prentice Hall. 2004.

Donnellan, Craig. *Genetic Modification (Issues)*. Independence Educational Publishers. 2004.

Friedberg, E.C. *DNA Repair*. New York: WH Freeman and Company. 1985.

George, E. F. Hall, M. A. and G.-J. De Klerk (eds). *Plant propagation by tissue culture. Volume 1. The background. 3rd edition* Springer, Dordrecht, 2008.

Gianessi, L. P., C. S. Silvers, S. Sankula, and J. E. Carpenter, "Plant Biotechnology: Current and Potential Impact for Improving Pest Management in U.S. Agriculture, an Analysis of 40 Case Studies", Washington, D.C.: National Center for Food and Agricultural Policy, 2002.

Goodman, D.C. *From Farming to Biotechnology: A Theory of Agro-industrial Development.* Oxford: Blackwell. 1987.

IPGRI. 2004. International Plant Genetic Resources Institute 2004). Promoting fonio production in West and Central Africa through germplasm management and improvement of post harvest technology. Final report.

Jain, S. Mohan, D. S. Brar, and B. S. Ahloowalia, eds., *Molecular Techniques in Crop Improvement*, Dordrecht, Boston, and London: Kluwer Academic Publishers, 2002.

James C. 2010. *Global Status of Commercialized biotech/GM Crops: 2009.* International Service for the Acquisition of Agri-biotech Applications (ISAAA) Briefs No. 41. Ithaca, NY.

James, C. Global review of commercialized genetically modified crops. ISAAA Brief Number 8. *International Service for the Acquisition of Agri-biotech Applications,* Ithaca, New York, USA. 1998.

Kloppenburg, J. R., Jr., *First the Seed: The Political Economy of Plant Biotechnology*, Cambridge: Cambridge University Press, 1988.

Morgan, Sally. *Superfoods: Genetic Modification of Foods (Science at the Edge).* Heinemann. 2003.

Murashige, T. "Plant propagation through tissue culture". *Annual Reviews Plant Physiology* 25: 135-166. 1974.

Regal, P. A brief history of biotechnology risk debates and policies in the United States. *Occasional paper.* Edmonds Institute, Edmonds, Washington, USA. 1999.

Rissler, J., and M. Mellon. *The ecological risks of engineered crops.* MIT Press, Cambridge, Massachusetts, USA. 1996.

Roelofs, D. et al. "Functional ecological genomics to demonstrate general and specific responses to abiotic stress." *Functional Ecology* 22: 8–18, 2008.

Smiley, Sophie. *Genetic Modification: Study Guide (Exploring the Issues).* Independence Educational Publishers. 2005.

Twyman, Richard M. "Molecular farming in plants: host systems and expression technology". *Science Direct*. 28 October 2003.

Varshney RK, Close TJ, Singh NK, Hoisington DA, Cook DR. 2009. Orphan legume crops enter the genomics era! *Current Opinion in Plant Biology* 12:202–210.

Villalobos, V.M., Ferreira, P. & Mora, A. "The use of biotechnology in the conservation of tropical germplasm". *Biotech. Adv.*, 9: pp 197-215. 1991.

Wang, W., Vinocur, B. and Altman, A. "Plant responses to drought, salinity and extreme temperatures towards genetic engineering for stress tolerance." *Planta* 218: 1-14, 2007.

Wilkins, C.P., Dodds, J.H. "The application of tissue culture techniques to plant genetic conservation". *Science Progress* 68: 281-307. 1982.

Williams JT, Haq N. 2002. *Global research on underutilized crops*. An assessment of current activities and prospects for enhanced cooperation. ICUC, Souhampton, UK.